Structure and Performance of Manufacturing in Kenya

Studies on the African Economies

General Editors: **Paul Collier** and **Jan Willem Gunning**
Published in association with the Centre for the Study of African Economies, University of Oxford

This important new series provides authoritative analyses of Africa's economies, their performance and future prospects. The focus will be on applying recent advances in economic theory to African economies to illuminate and analyse the recent processes of economic reform and future challenges facing Africa. The books, published in association with the Centre for the Study of African Economies, will bring together top scholars from universities and international organisations across the world.

Titles include:

Arne Bigsten and Steve Kayizzi-Mugerwa
CRISIS, ADJUSTMENT AND GROWTH IN UGANDA
A Study of Adaptation in an African Economy

Arne Bigsten and Peter Kimuyu (*editors*)
STRUCTURE AND PERFORMANCE OF MANUFACTURING IN KENYA

Paul Collier and Catherine Pattillo (*editors*)
INVESTMENT AND RISK IN AFRICA

Paul Glewwe
THE ECONOMICS OF SCHOOL QUALITY INVESTMENTS IN DEVELOPING COUNTRIES
An Empirical Study of Ghana

Jan Willem Gunning and Remco Oostendorp (*editors*)
INDUSTRIAL CHANGE IN AFRICA
Zimbabwean Firms under Structural Adjustment

Carolyn Jenkins and John Knight
THE ECONOMIC DECLINE OF ZIMBABWE
Neither Growth nor Equity

John Knight and Carolyn Jenkins
ECONOMIC POLICIES AND OUTCOMES IN ZIMBABWE
Lessons for South Africa

Jo Ann Paulson (*editor*)
AFRICAN ECONOMIES IN TRANSITION
Volume 1: The Changing Role of the State
Volume 2: The Reform Experience

Structure and Performance of Manufacturing in Kenya

Edited by

Arne Bigsten

and

Peter Kimuyu

palgrave

First published 2002 by
PALGRAVE
Houndmills, Basingstoke, Hampshire RG21 6XS and
175 Fifth Avenue, New York, N. Y. 10010
Companies and representatives throughout the world

PALGRAVE is the new global academic imprint of St. Martin's Press LLC Scholarly and Reference Division and Palgrave Publishers Ltd (formerly Macmillan Press Ltd).

ISBN 0–333–97028–4

This book is printed on paper suitable for recycling and made from fully managed and sustained forest sources.

A catalogue record for this book is available from the British Library.

Library of Congress Cataloging-in-Publication Data
Structure and performance of manufacturing in Kenya / edited by
Arne Bigsten and Peter Kimuyu.
p. cm — (Studies on the African economies)
ISBN 0–333–97028–4 (cloth : alk. paper)
1. Manufacturing industries—Kenya. I. Bigsten, Arne. II. Kimuyu, Peter. III. Series.
HD9737.K4 S77 2001
338.4'767'096762—dc21 2001045150

10 9 8 7 6 5 4 3 2 1
11 10 09 08 07 06 05 04 03 02

Printed and bound in Great Britain by
Antony Rowe Ltd, Chippenham, Wiltshire

Contents

List of Figures and Tables

Figures

Tables

Foreword

There is no more appropriate time to publish a book on the state of manufacturing in Africa than now. African governments are frantically scratching for development ideas that can make a difference to the prospects of the African people in a continent where development efforts have not borne much fruit. Poverty levels are high, so that poverty reduction has become a buzz word in circles where Africa's development is discussed. It is important to begin examining the relative contributions of different sectors in empowering Africans so that emphasis can be appropriately placed.

Although the link between industrialisation and poverty reduction has yet to be established, Kenya's current development strategy is about making Kenya an industrialising country in the near future. This strategy results from a realisation that the country has reached a development threshold were agriculture, the country's back bone, can no longer on its own generate the necessary momentum. Although manufacturing activities have gone on for a long time in most of Africa, the contribution of these activities to gross domestic product (GDP) remains very low. Structural changes that have taken place in most of Africa seem to have by-passed the manufacturing sector, elevating the importance of the service sector which has become the main rival to agriculture. This is therefore an appropriate juncture to explore the salient features of manufacturing and its outcomes in order to increase the stock of knowledge from which interventions for leveraging African development can be extracted. This is what the book is all about.

Manufacturing is of course uniquely able to add value and create linkages across different sectors of an economy. Obviously, manufacturing requires institutions that are more efficacious than those that support general commerce. In the past, it was possible to argue that a country's proportions of resources and skills define the threshold within which manufacturing can take place. However, globalisation has expanded the space for fine-tuning commodity and value chains which permit exploitation of global cost and convenience advantages in the production of products and product components. Evidently, manufacturers that operate in costly business environments face penalties that reduce their ability to penetrate and maintain a presence in the regional and global arena. The quality of such an environment determines the ability of enterprises to respond to policy reforms and secure market shares in regional and international markets. The environment also determines the operational threshold of factor productivity and return on investment. Shedding light on the environment in which manufacturing activities are carried out is therefore important in explaining the dominant business philosophy, entrepreneurs' response and enterprises performance.

In this book, the authors exploit unique data sets to investigate different aspects of manufacturing during a period of rapid reform. The book is therefore situated in a rare historical juncture that permits an investigation of the dynamics of firm response to changes in the incentive structure. Not only does the book increase the stock of knowledge on the differential impact of size, sector, location, formality and ethnicity on performance, it contains entrepreneurs' ideas about what the future of the manufacturing sector could look like. But it is perhaps the size factor that pops up in the different sections of the analysis and must inform evolution of industrial policy in Africa. The balance of the evidence is that the small manufacturers perform rather badly on most scores, so that small may not be beautiful after all.

This study was done within the World Bank's Regional Enterprise Development Programme (RPED). The RPED project was organized by Tyler Biggs, who was assisted by Manju Kedia, Melanie Mbuyi and Alfred Robinson. Tyler was responsible for designing of the questionnaire and took part in the pilot stage of the study. We are very grateful for his input. The three waves of the Kenyan surveys were organised by the Economics Departments of Gothenburg and Nairobi Universities in collaboration with the Kenya Association of Manufacturers. Valuable help was also received from the Central Bureau of Statistics. The interviews were carried out by a team that included Peter Kimuyu, Gerrishon Ikiara, Mohamud Jama, Mutsembi Manundu, Wafula Masai, Walter Ochoro, Njuguna Ndung'u, Muhammud Jama, N.E.O. Ambitho, E.M. Kabuu, Simon Ihiga, Paul Mwake, Anders Gerdin, Anders Isaksson, Jörgen Levin, Steve Mugerwa, Clas Wihlborg, Lennart Hjalmarsson, and Arne Bigsten. The data entry was done by Domisiano Mwabu Kirii and Geoffrey Opicha Okallo. The organisation of the data set was then done by Renato Aguilar. Finally, we would like to acknowledge that our greatest debt of gratitude is to the firms interviewed for their patience, endurance and collaboration.

ARNE BIGSTEN
PETER KIMUYU

Notes on the Contributors

Renato Aguilar is Associate Professor at the Department of Economics at Göteborg University. He has worked in various countries in Africa, mostly Portuguese-speaking Africa, and Latin America, primarily Chile, Nicaragua, Panama and Cuba. His recent publications are about Angola (in English) and Cuba's agriculture and poverty (in Spanish).

Arne Bigsten is Professor of Development Economics at Göteborg University. He has done extensive research on African economic development, with regard to income distribution and poverty, industrial and rural development, trade, aid and economic policy. He is currently Vice Chair of the EU Development Network.

Mats Granér is a graduate student in economics at Göteborg University. His research is mainly concerned with the link between trade openness and productivity in developing countries. He is at present working at the Swedish Ministry of Industry, Employment and Communications.

Lennart Hjalmarsson is August Rohss Professor in Economics at Göteborg University. His main research fields are industrial and energy economics. He has served as Chairman and member on several government panels, especially on energy policy issues in Sweden. He has also served on the board of several scientific councils and scientific journals and is a member of the Royal Swedish Academy of Engineering Sciences.

Gerrishon K. Ikiara is a Senior Lecturer at the Institute of Diplomacy and International Relations, University of Nairobi. Previously, he was Senior Lecturer in the Department of Economics, University of Nairobi. He has researched and published extensively on various aspects of the Kenyan Economy. He is the co-editor *of Industrialisation in Kenya: in search of a strategy* (Heinemann (K), 1988) and *The Industrialisation Dilemma* (Heinemann (K), 1991).

Anders Isaksson obtained his PhD in economics at Göteborg University in 1997. His research interests are in financial economics, international trade, and macroeconomics, all with a particular focus on developing countries. Anders Isaksson is now at the United Nations Industrial Development Organization (UNIDO).

Steve Kayizzi-Mugerwa is an Associate Professor in Development Economics, Göteborg University. His PhD dissertation was on shocks and economic reform in Zambia. He has worked at the African Development Bank and currently directs a project on African institutions and reform ownership at WIDER, United Nations University, in Helsinki. Besides research collaboration with a number of African universities, he has also carried out consultancy work for the World Bank, the Swedish Government and the OECD Development Centre. His publications include more than 20 articles in international journals and co-authored and edited books on economic development in Africa.

Peter Kimuyu is the Executive Director of the Institute of Policy Analysis and Research in Nairobi, Kenya, and the immediate former head of the Economics Department at the University of Nairobi. He has been a visiting fellow at Göteborg University and Williams College, Massachusetts. He has published on energy, industry, entrepreneurship and rural development.

Jörgen Levin has a PhD from Göteborg University and is now a Lecturer at the Department of Economics, Örebro University. He has worked on various issues related to structural adjustment in Kenya. He is currently working on a project analysing economic growth, fiscal policies and poverty reduction in Tanzania.

Karl Lundvall, currently working with the Swedish Competition Authority, has a PhD in development economics from Göteborg University. He also has working experience in debt management from the Ministry of Finance in Dar es Salaam. His research has focused on efficiency and productivity analyses within African manufacturing.

Damiano Kulundu Manda is an analyst at the social sector division of the Kenya Institute of Public Policy Research and Analysis (KIPPRA). He holds a doctorate in economics from the Göteborg University and is now on leave from the University of Nairobi, where he is a Lecturer in Economics. His research interests cover labour markets and poverty, as well as health and education policy issues.

Mutsembi Manundu is a Lecturer in Economics in the University of Nairobi. He has postgraduate qualifications from the University of Michigan. His research interests include international and development economics, industrialisation and trade, and environmental issues.

Wafula Masai holds a doctorate in economics from the University of Grenoble, France. He is a senior lecturer in, and current chairman of, the Economics Department at the University of Nairobi. He is also the chairman of the Academic Board of the African Economic Research Consortium that supports a regional collaborative masters programme in Anglophone

Africa. His research interests include economic policy analysis, operations analysis, industrial economics and human resource development.

Njuguna S. Ndung'u is a Principal Analyst at KIPPRA and Senior Lecturer in the Department of Economics, University of Nairobi. He has a PhD in economics from the Göteborg University. He has published widely on macroeconomic issues, economic growth, regional integration, pro-poor growth strategies and external debt problems. He is a researcher and resource person with the Africa Economic Research Consortium and has been a visiting scholar in the International Monetary Fund (IMF) Research Department, the United Nations Economic Commission for Africa and the World Bank, among others.

Walter Ochoro is a Lecturer in Economics at the University of Nairobi. His main areas of research interest include mathematical economics, econometrics, international trade, industrial and agricultural economics, economics of technological change and diffusion of technologies, technical and allocative efficiency efficiencies, productivity and productivity change, stochastic and deterministic quantitative modeling.

Måns Söderbom has a PhD from Göteborg University and is now the UNIDO Research Fellow at the Centre for the Study of African Economies, University of Oxford. The focus of his research is the microeconomics of African manufacturing, particularly investment behaviour, the determinants of exports, production efficiency, the role of human capital and behaviour under risk.

Clas Wihlborg is Professor of Finance at the Copenhagen Business School and Professor of Banking and Financial Economics at Göteborg University. He has published books and articles in the areas of international finance, macroeconomics and financial institutions. During the last decade he has focused his research on financial institutions in emerging market economies. He is a member of the Royal Swedish Academy of Engineering Sciences, and the European Shadow Financial Regulatory Committee.

List of Abbreviations

CMA	Capital Markets Authority
COMESA	Common Market for Eastern and Southern Africa
CPI	consumer price index
CRS	constant returns to scale
DEA	Data Envelopment Analysis
EAC	East African Community
EPPO	Export Promotion Programme Office
EPZ	Export Processing Zone
ERP	effective rates of protection
ESAF	Enhanced Structural Adjustment Facility
FE	fixed effects
ForexCs	Foreign Exchange Certificates
IDA	International Development Association
IFC	International Finance Corporation
ILO	International Labour Organisation
IMF	International Monetary Fund
KPA	Kenya Ports Authority
KPTC	Kenya Posts and Telecommunications Corporation
MLE	Maximum Likelihood Estimation
MNCs	multinational corporations
NBFI	Non-Bank Financial Institution
NOCs	No-Objection Certificates
NSE	Nairobi Stock Exchange
OLS	ordinary least squares
PTA	preferential trade area
RE	random effects
RPED	Regional Programme of Enterprise Development
TFP	total factor productivity
UNIDO	United Nations Industrial Development Organisation
VAT	value-added tax

Part I

Introduction and the Regional Programme of Enterprise Development Survey

1
Introduction

Arne Bigsten and Peter Kimuyu

When African countries gained political independence their governments had ambitious plans to modernise their economies. It was generally felt that the private sector was inadequate for the task and that extensive government intervention and public ownership of firms were required. The strategy chosen was one of import substitution industrialisation. There was extensive investment in state corporations, which were generally less efficient than their private counterparts. The problem of achieving an efficient allocation of resources was further aggravated by highly distorted price structures, particularly overvalued exchange rates.

The chosen development model came under severe pressure as African economies were severely affected by the first oil shock in 1973, but most of them chose to postpone adjustment. However, when commodity prices tumbled at the end of the 1970s and the era of cheap credit came to an end the situation became untenable for many countries, which then embarked on adjustment efforts. Adjustment programmes have generally combined macroeconomic stabilisation and structural reforms. The aim has been to put a more liberal market system into place. It has included measures directed at the credit markets, factor markets, external and internal trade and the foreign exchange market. Adjustment also encompassed institutional reforms, particularly measures to slim down the public sector. There was also a strong emphasis on the development of an environment that could enable private business to expand more rapidly.

The initial stages of the adjustment efforts generally combined contraction of domestic demand with attempts to expand external markets via depreciations of the currency. The former had an immediate impact, while the export markets have been slower to respond. One would therefore expect to see an initial slowdown in growth, followed by a gradual recovery. This is also the pattern that was observed in several countries in the 1980s. Over the whole decade, close to half of the countries in Africa saw per capita incomes increase, while the remainder experienced further declines. In the early 1990s Africa was also affected by the global recession and falling terms of trade, which meant that only a third of the countries could continue to increase per capita incomes during the first half of the decade.

In the 1990s there were some positive signs for exports, which essentially derive from the private sector (Kirkpatrick and Weiss, 1995). In spite of the adverse international conditions, sub-Saharan Africa managed to expand merchandise exports by 2 per cent per year, which is faster than in the 1980s. When adjustment started and demand had to contract, investment was the main casualty. For sub-Saharan Africa as a whole the share of GDP going to investment after that has remained fairly constant at 16–17 per cent since the 1980s. It is noteworthy that the share of private investment has generally remained low.

Domestic resource mobilisation has been poor in many African countries. While Africa invests almost the same fraction of GDP in the public sector as other developing regions, the share of GDP going to private investment is low. Private capital flows, which are the dominant part of capital flows to less developed countries today, are largely passing Africa by (Jaspersen, Aylward and Sumlinski, 1995; Lesink and White, 1998). The private sector has thus not yet taken off in the majority of African countries, although the decline of the 1980s has been halted and in some cases reversed. The main conclusion about the impact of structural adjustment measures drawn by the World Bank (1994) was that those countries with large improvements in the macroeconomic framework (exchange rate policy, inflation, budget deficits) have done better in terms of growth than those without such improvement. Although improvements in the various macroeconomic indicators relating to fiscal, monetary and exchange rate policies are necessary for recovery, they only prepare the ground for improvements in variables necessary for long-term development.

Sustainable higher growth rates require higher private savings and investments. Still, savings and investments in sub-Saharan Africa are significantly lower than in other regions. The policies need to be such that they stimulate both domestic savings and investment at the same time as they attract foreign investment.

In an extensive review of adjustment experiences in Africa 1986–93, Hadjimichael and Ghura (1995) found that although the general picture was rather bleak, there were extensive differences in performance among the sub-Saharan economies. Preserving international competitiveness and reducing impediments to private sector activity while striving for macroeconomic stabilisation was found to be crucial. Sustained adjusters and countries with low macroeconomic imbalances were doing better than others. It is clear that investment has a positive impact on growth, but so far it has been hard to discern a positive link from adjustment to investment. There is a need for a process of confidence building as well as further public investment in infrastructure to elicit private sector investment responses.

Before the adjustment process started the industrial sector was heavily protected and benefited from various types of subsidies. It had, for example, favoured access to foreign exchange at an overvalued exchange rate. When these advantages disappear, it is inevitable that many of the firms that thrived behind these protective walls will fail. Since adjustment is about changing relative prices, other sectors and activities that benefit from improved relative prices should prosper.

While there are signs of economic recovery in several African countries, there are also countries where incomes have fallen further. Some countries have managed to create an environment conducive to private sector growth, while in others a lot remains to be done. Private sector expansion is necessary for economic revival, and this requires private sector investment. There has been some progress, but generally it seems to take a long time to build up investor confidence. It may therefore be too early to say that growth led by the private sector is generally taking hold, although there are promising developments in some of the countries.

1.1 The Kenyan case

Kenya is in many ways a typical example of the pattern of development just described. From independence in 1963 up to the first oil price shock of 1973/74, Kenya experienced relatively high growth rates. The oil shock was followed by a series of other disturbances, which further destabilised the economy. By the mid-1980s the economy had regained a measure of stability, and the years 1986–90 saw fairly high and stable growth, on average 5 per cent per year. Then the economy stagnated again, resulting in falling per capita incomes until the mid-1990s. This was partly due to international events, partly to slippage in macroeconomic management and donor reaction to it.

In spite of some progress in macroeconomic stabilisation during the second half of the 1980s, exemplified by the move towards a more competitive exchange rate, the impact on industry, particularly industrial exports, was limited. The slow growth of private industrial investment was a major setback. It is broadly acknowledged that a number of structural and institutional variables, quite apart from those of a macroeconomic nature such as the exchange rate, impinge on industrial development. Thus, while macroeconomic stability is still very important, the policy debate has shifted to sectoral and microeconomic issues such as investment policy, business support systems, public sector divestment and infrastructure development. To be able to move towards growth rates that are both higher and better sustained than those experienced so far, investment and productivity will need to be stimulated. Kenya's public sector investments have, on average, performed poorly. It is necessary to develop an environment in which private investment can thrive.

This study analyses factors that affect the efficiency of business, and thereby investment, decisions in Kenya. It is a study of the enabling environment for private industry. The study has shed light on (1) the structure of economic incentives and the regulatory policies that constrain the efficient allocation of labour and capital; (2) the different components of the transaction cost environment (e.g., information flows, property rights, legal structures, and infrastructure); (3) factors that inhibit the acquisition of technological capability; and (4) the operation of business support agents involved in finance, the supply of technical training services, and the provision of market information. Our analysis of data on a panel of manufacturing firms provides insights into both the effects of the current business environment and those of policy changes on business and investment decisions. It also permits an expansion of the dynamics of enterprise activity – the rate of entry, growth and exit – in Kenya.

The presentation proceeds as follows: Chapter 2 puts this study into context by providing a historical review of manufacturing and the relevant policy-making in Kenya. Chapter 3 then presents the surveys on which the study is based. These three chapters form the introductory part of the book. The second part of the book is about the structure of Kenyan manufacturing. Chapter 4 considers the character of the firms and product markets, Chapter 5 looks at the structure of the labour market, and Chapter 6 investigates the financial markets. The third part analyses investment behaviour of Kenyan manufacturing firms, which is a key to long-term growth, as has just been indicated. In Part IV we investigate the determinants and changes in productivity and efficiency in Chapter 8, before looking at export performance, which is arguably an essential aspect if the economy is to take off into a pattern of sustainable growth. Chapter 10 completes Part IV by investigating interfirm differences in growth. Finally, in Part V we first look at the impact of macroeconomic policy reforms on the manufacturing sector (Chapter 11) and then consider the impact of infrastructure on the performance of the sector (Chapter 12). Chapter 13 concludes the book.

References

Hadjimichael, M.T. and Ghura, D. (1995), *Public Policies and Private Savings and Investment in Sub-Saharan Africa: An Empirical Investigation*, Africa Department, International Monetary Fund (IMF) Working Paper, IMF, Washington, DC.

Jaspersen, F.Z., Aylward, A.H. and Sumlinski, M.A. (1995), *Trends in Private Investment in Developing Countries. Statistics for 1970–94*, Discussion Paper no. 28, International Finance Corporation, Washington, DC.

Kirkpatrick, C. and Weiss, J. (1995), 'Trade Policy Reforms and Performance in Africa in the 1980s', *Journal of Modern African Studies* 33(2).

Lesink, R. and White, H. (1998), 'Does the Revival of International Private Capital Flows Mean the End of Aid? An Analysis of Developing Countries' Access to Private Capital', *World Development* 26(7).

World Bank (1994), *Adjustment in Africa*, Oxford University Press, New York.

2
History and Policy of Manufacturing in Kenya

Arne Bigsten

2.1 Introduction

The purpose of this chapter is to provide a background to the subsequent chapters, which are devoted to the analysis of the Regional Programme of Enterprise Development (RPED) data covering the development of the manufacturing sector in Kenya between 1992 and 1995. The Kenya survey was part of a multicountry World Bank study, and was undertaken jointly by the Economics Departments of Göteborg University and Nairobi University and the Kenya Association of Manufacturers. The survey is described in detail in Chapter 3.

In this background chapter we review how the aspects of trade, industrial and macroeconomic policies most relevant to the manufacturing sector have evolved and how the manufacturing sector in Kenya has changed. We start in section 2.2 with a few brief comments on the colonial period, and then in section 2.3 we consider the period from independence to 1970, a period of moderate import substitution. In section 2.4 we discuss the developments during the 1970s, which was the most intensive period in terms of import substitution and also a period of fairly rapid manufacturing growth. Section 2.5 focuses on the shift of policy away from import substitution towards export promotion during the period of structural adjustment in the 1980s and 1990s, while we discuss the shortcomings of the process of policy-making and the conflicts with the donor community in section 2.6. In section 2.7 we review the impact of the reforms on growth and investment, particularly in the manufacturing sector. Section 2.8 gives some facts about long term productivity growth in Kenyan manufacturing, while section 2.9 provides concluding remarks.

2.2 The colonial period

The modern manufacturing sector in Kenya has its beginnings in the colonial period, when an economic infrastructure linking Kenya to the world market emerged. Import–export houses were established in Kenya as early as 1905. At the beginning these were involved in the export of primary commodities and the import of consumer goods, primarily for the settlers.

In the interwar period some manufacturing industries were established, mainly producing consumer goods for the domestic or East African market. Most of them were processing agricultural goods, although some other types of industries were also established at this time: for example, The Magadi Soda Company, East African Power and Lighting, and East African Portland Cement. Many of the latter were subsidiaries of major multinational firms. Since the colonial authorities at this time excluded the Asians from farming in the White Highlands, they moved into urban centres and invested in small-scale manufacturing activities along with shops and real estate. Practically all firms founded at the time were established without direct state support.

After the Second World War Britain introduced tariffs in Kenya to protect their colonial markets against foreign competition, and a range of British firms invested in East Africa (Nyongo, 1991). The market was limited and allowed only small enterprises, which would have to grow more or less at the same rate as the domestic market unless they could expand into exports. Given the shortages of local capital and skilled management, penetrating export markets seemed unlikely to happen on a large scale. It was assumed that for an extended period of time these resources would continue to be supplied from the outside and that the market would be confined to Kenya and to some extent the neighbouring countries. Although there was some concern about the long-term effects of protectionist policies, they were kept in place.[1]

2.3 The early independence years, 1963–70: moderate import substitution

There was a slowdown in industrial investment and considerable transfer of capital out of Kenya in the period of transition to independence in 1963. To restore investor confidence and halt the capital outflow, the new government passed the Foreign Investment Protection Act in 1964. On the basis of this firms were given certificates that guaranteed them the right to repatriate profits, loans and interest on their loans, and part of the proceeds from sales of assets. The government also guaranteed that the firms would not be nationalised. The following year the government published its most famous White Paper, *Sessional Paper no. 10 of 1965 on African Socialism and its Application to Planning in Kenya* (Kenya, 1965). In this paper, the government spelled out its ambition to develop a free market economy, in which foreign investors were welcome. Unlike some of its neighbours, the government rejected the idea of extensive nationalisations.

The first years after independence saw a continuation and eventually a deepening of the import-substitution industrialisation policy, since this at the time was seen by most as the natural course of industrialisation (see, e.g., van Arkadie, 1964). An interministerial Industrial Protection Committee was set up to consider protection issues. It considered applications from firms seeking modifications of the structure of tariffs, quotas, duty drawbacks and import licensing (Gerdin, 1997). A Foreign Exchange Allocation Committee was set up to administer import quotas on imports for which limited quotas had been established. The quotas were typically set equal to the difference between predicted local production and local demand. With increasing foreign exchange problems in the early 1970s quotas were tightened, and at times applied to ban imports of luxury goods. The committee was an important part of the import substitution policy-making, but became redundant with the liberalisation of the foreign exchange market in 1993/94.

By 1964 investor confidence in Kenya was largely restored, and foreign capital started to flow back into Kenya. The foreign investors began to focus on manufacturing rather than agriculture, and came to dominate in areas such as footwear, leather, rubber, petroleum, industrial chemicals, paints, soft drinks, cement and metal products (Gerdin, 1997). Some long-established foreign trading and service companies also shifted into manufacturing as the government increasingly tried to restrict trade to Africans. Immediately after independence much of the foreign investment capital was simply transferred out of agriculture, and a few years later, new foreign capital was flowing in as industries such as Union Carbide, Firestone, United Steel, Del Monte, Schweppes and Lonrho started producing for the domestic and regional markets. The bulk of foreign investment went into import-substituting industries, and the established firms often enjoyed a virtual monopoly. The effective rates of protection were very substantial, particularly for consumer goods industries (Eglin, 1978). There were also some joint ventures between foreign companies and government-owned development corporations (accounting for 9 per cent of new foreign investment between 1964 and 1970), but the results of those were often poor.

The policy of indigenisation was an important contributor to the spread and structure of manufacturing. At independence, Kenya's industrial sector was dominated by firms owned by Asians and Europeans. Although the industry had reached a relatively sophisticated level *vis-à-vis* the neighbouring countries, it was felt that Africans had not benefited sufficiently. A substantial amount of legislation was put in place in an attempt to redress this legacy. For instance, to further African involvement in economic activity the government introduced the Trade Licensing Act of 1967, which excluded non-citizens from certain business activities, and the Immigration Act of the same year, which restricted employment of non-citizens. A Kenyanisation of Personnel Bureau was established to Africanise the top posts in the civil service and the parastatals. Work permits were introduced to regulate the number of foreign workers in the private sector.

Another instrument in the hands of the authorities was industrial licensing, through which policy-makers could, at least potentially, discriminate in favour of Africans. Ironically, entry restrictions in retail and wholesale trade led many Asians into the light manufacturing industry, where they are clearly dominant today.

Policies to encourage indigenous businessmen's participation in Kenyan industry have often been half-hearted and borne little fruit. A small number of local entrepreneurs, exploiting either family or business ties with the country's leadership, have been able to acquire multiple directorships and shares in some ongoing manufacturing concerns. By contrast, a number of up and coming indigenous industrialists in the post-independence period appear to have fallen out and/or provoked the discomfort of the leaders and therefore experienced a sudden change of fortunes.

Asian investments in manufacturing accelerated from the 1950s, although they were held back by political hostility and discrimination that made it hard, for example, for Asians to obtain citizenship. Capital held by non-Kenyan but long-resident Asians did not qualify under the terms of the Foreign Investment Protection Act, which meant that they were unable to remit profits legally (Gerdin, 1997). Given their insecure position, the incentive to remit was strong and other ways of transferring money were obviously found.

In 1968 the government established the New Projects Committee to review applications from multinational corporations (MNCs) to invest in Kenya and negotiate with the foreign investors on protection levels, duty exemptions, levels of government shareholding, Africanisation and other matters. The committee became redundant after the establishment of the Investment Advisory Centre in 1983 and was abolished thereafter. Over the years, the New Projects Committee had been exposed to considerable political interference (Swainson, 1980), a feature which has characterised the Kenyan economy since independence. Straddling both politics and business has been common in Kenya and is likely to have had considerable distortionary effects on the economy, reducing growth (see Bigsten and Moene, 1996).

During 1964–70, the manufacturing sector experienced a rapid import-substitution related expansion, in which textiles and garments, food, beverages and tobacco were the leading sectors. This expansion was thus primarily based on the policy measures outlined above: that is, a combination of tariffs and import quotas supplemented by foreign exchange allocation measures, supplemented by the effects of an overvalued exchange rate. Interest rate and credit policies were geared towards subsidising producers of manufactures. In spite of these distortionary policies there was increasing exportation of manufacturing products during this period, primarily to the East African region.

Table 2.1 shows how the structure of the modern sector of the Kenyan economy changed during the first 30 years of independence. We note that

Table 2.1 Sectoral GDP as shares of monetary sector GDP in constant 1982 prices for 1964–95 and at current 1995 prices

	1964	*1970*	*1975*	*1980*	*1985*	*1990*	*1995*	*Current prices 1995*
Agriculture	0.414	0.376	0.361	0.328	0.302	0.302	0.271	0.294
Mining	0.003	0.004	0.004	0.003	0.003	0.003	0.003	0.002
Manufacturing	0.079	0.084	0.104	0.130	0.132	0.136	0.138	0.095
Electricity	0.005	0.005	0.007	0.008	0.009	0.011	0.010	0.009
Building	0.042	0.051	0.047	0.047	0.034	0.033	0.025	0.039
Wholesale	0.172	0.156	0.121	0.118	0.111	0.113	0.120	0.157
Transport	0.051	0.059	0.053	0.055	0.064	0.060	0.062	0.074
Finance	0.034	0.047	0.052	0.063	0.076	0.081	0.100	0.115
Ownership of dwellings	0.074	0.068	0.067	0.061	0.059	0.056	0.056	0.040
Community services	0.126	0.150	0.185	0.187	0.198	0.206	0.216	0.177

Sources: Gerdin (1997), p. 40; *Economic Survey* (1998), p. 24.

during the 1964–70 period the structure of the economy changed relatively little, although there was a modest increase in the share of manufacturing. The big expansion of manufacturing was to come in the 1970s. We will return to this table in our discussion below.

2.4 The turbulent 1970s: intensified import substitution

In 1970–1 Kenya was hit by an economic crisis as foreign exchange reserves dwindled due partly to too expansionary a fiscal policy (Bevan *et al.*, 1990). The government reacted to the crisis by tightening its administrative control on the economy. In 1971 it set up the Capital Issues Committee to vet stock issues in order to cut down on capital outflows and to encourage local ownership. Tariff levels went up and import licensing became more restrictive. Since this was expected to have price effects, the government also tightened price controls in 1972. The price controls were also to serve as a basis for an income policy, and the first wage guidelines were issued in 1973. In 1972 the government had introduced a sales tax which in theory was neutral between domestic producers and foreign competitors. However, it would seem that domestic producers did not pay their dues, thus again receiving further implicit protection (Gerdin, 1997). In the early 1970s the government also introduced the highly protective No-Objection Certificates (NOCs), which ensured that importers made sure that goods similar to those they wished to import were not produced in Kenya. The NOCs were discontinued ten years after introduction.

These extensive interventions in the early 1970s further reduced domestic competition and shifted incentives against export production. That there really was a strong trend towards import substitution and inward orientation in manufacturing is shown by the fact that the share of domestic production in total supply of manufactures increased from about 50 per cent in the mid-1960s to nearly 75 per cent in the mid-1980s. At the same time export production shrank from as much as 40 per cent of the value of manufacturing output in 1964 to about 10 per cent in 1984 (Sharpley and Lewis, 1990).

The policy pursued not only distorted incentives between manufacturing exports and imports, but also implied a strong subsidy of manufacturing at the expense of the rest of the economy, notably agriculture. Sharpley and Lewis (1990) estimate that the increased subsidy going to manufacturing due to protection during 1964–84 was equivalent to 28 per cent of the increase in agricultural output. This system was sustainable as long as agriculture was growing rapidly, but once agricultural growth slowed down, the basis for this strategy of manufacturing growth was also being eroded. Yet the exploitation of agriculture was probably less in Kenya than in several other African countries, because the ruling elite was largely agriculturally based.

Import-substituting manufacturing in Kenya grew rapidly during the 1970s and the share of manufacturing in the modern sector of the economy increased from 8.4 per cent to 13 per cent at constant prices during the decade (see Table 2.1). There was also a considerable diversification of the manufacturing sector in Kenya: for example, in areas such as plastics, pharmaceuticals, electric cables, paper, vehicle assembly, industrial gases, rubber and batteries (Coughlin, 1990).

However, at the end of the 1970s the scope for further consumer goods import substitution based on simple assembly or compounding of imported materials seemed to be exhausted.[2] Domestic and foreign demand slackened and structural problems set in. Demand growth slowed down with the deceleration in agricultural activity and the loss of access to regional markets. Moreover, the inward-looking policies pursued at the time reduced the competitiveness of Kenyan products in the export markets. The export compensation scheme that was introduced in 1975 was not sufficient to offset this overall bias against manufactured exports.

The import-substitution policies did not emphasise the production of intermediates and capital goods, which meant that there was a rapid increase in demand for foreign exchange from the expanding consumer goods industries. Given the overvaluation of the exchange rate, imported capital goods were relatively cheap, which further reduced incentives to start domestically competing production. The interviews undertaken within the RPED project have shown clearly that Kenyan firms undertake neither technological adaptation nor research and development. The import substitution industrialisation policies which were pursued actually encouraged the importation of whole plants, by having very low or no tariffs on machinery and high tariffs on engineering raw materials (Juma, 1991). Suitably skilled manpower has therefore not emerged in Kenya.

There was a slowdown in MNC investment in manufacturing in Kenya from the mid-1970s. The dominance of the multinationals in manufacturing was thus reduced during the 1960s and 1970s, mainly by the expansion of Kenyan Asian firms and the parastatals. In the 1980s capital outflows for equity and business debt repatriations were three times the inflow (Vaitsos, 1991, p. 9). The inflow of equity capital in the 1970s had been mainly into textiles, but this was reversed in the 1980s when food, beverages and agro-industries once again took the lead in manufacturing. However, even more capital went into non-manufacturing sectors, such as banking and finance.

Vaitsos (1991, p. 25) notes that foreign firms have invested in Kenya when demand has been buoyant. Therefore, to attract foreign capital the economy must have growth potential. Providing incentives for producers is not enough unless there is demand for the output. Of course, if the environment was favourable enough, foreign firms might locate there and use Kenya as an export base. This has, so far, only happened on a very limited scale.

An interesting feature of Kenyan industrialisation is the proliferation of the 'open air' (*Jua Kali*) manufacturing activities, not only in the major cities but also in the countryside. These activities were first brought into the limelight by the ILO report of 1972 (ILO, 1972). Following the various economic crises and the stagnation of the formal sector, the informal sector has expanded rapidly. Informal production activities comprise small undertakings employing, in most cases, fewer than 10 workers, are often unregistered, and use production methods which require limited specialisation and management capacity. A large proportion of the output of these firms is directed towards satisfying basic needs, namely the provision of consumer goods and services for low-income households. Examples of the commodities produced include clothing, furniture, foodstuffs and metal utensils. Many factors have contributed to the rapid growth of the informal sector: the capital needed to embark on production is minimal; often labour is supplied by family members; producers are situated near the market; and due to their 'informality' many producers do manage to avoid numerous taxes that are imposed on formal firms.

It was hoped that the informal sector would be the springboard for an indigenous form of industrialisation: all the ingredients seemed to be there. There was a market for the goods produced in the sector, and there seemed to be abundant opportunities for the spread of the learning-by-doing effects. The derived savings would lead to further investment. There has also been a substantial expansion of informal employment over the years, and by the mid-1990s there were more people in informal than in formal employment. However, many producers seem to see their activities mainly as a way of generating funds to enable them to embark on other activities (e.g., agriculture, or the maintenance of their families). The firms seldom seem to remain viable beyond the active life of the proprietors. The turnover, in terms of establishments, is thus substantial. Moreover, owing to the semi-legal character of the activities, say with respect to tax avoidance, informal firms seem to prefer not to become too large. As a result, there is a lot of horizontal expansion.[3] King and Aboudha (1991) provide a more optimistic view of the potential of the informal sector. Comparing a sample of entrepreneurs revisited 18 years after the first interviews in the early 1970s, the authors noted substantial mechanisation and expansion of activities. There was also a tendency to move into more permanent premises.

From the 1970s, one sees a clear pattern emerges in the structure of the manufacturing sector in Kenya: the large industries are dominated by parastatals and multinationals, the small to medium-sized firms by Asians and the micro and informal sector ones by Africans.

2.5 Structural adjustment in the 1980s and 1990s: transition towards export promotion

By 1980 Kenya had essentially undertaken the import substitutions that policy-makers had envisaged. Consumer goods imports were very low, and the government began to realise that there was no more scope for further import substitution of consumer goods. It also saw the distortionary effects in the form of, for example, reduced competition and discrimination of other exportables producers. There was also a concern about the low capacity utilisation of the industry, and increasingly the policy-makers focused on efficiency rather than protection issues (Gerdin, 1997).[4]

An important turning point in Kenya's industrial policy came with the introduction of the structural adjustment programmes from the beginning of the 1980s to deal with the country's macroeconomic imbalance as well as structural crises. We will in this section discuss the various policy changes that were of relevance to the manufacturing firms.

2.5.1 Tariff reforms

Their main thrust was to effect a shift from a highly protective import-substitution strategy to industrial policies which would lead to increased use of local resources, greater employment creation and the encouragement of exports. The three main objectives of the industrial sector reform programme were: (1) to improve efficiency; (2) to stimulate private investment; and (3) to increase the sector's net foreign exchange earnings while diversifying its resources. In addition, and as part of the macroeconomic framework that underpinned the programmes, the Government of Kenya undertook to pursue a flexible exchange rate policy. This meant that the real exchange rate depreciated from about 1984, but generally the implementation of the reform measures was half-hearted during the first half of the 1980s.

The prevalence of quantitative import restrictions meant that there was very little competition from imports. From the mid-1980s a process of revision of trade policy started (see Tables 2.2 and 2.3). The number of different *ad valorem* tariff rates was reduced and a range of other specific provisions were abolished. By 1996/97 the top tariff rate had been reduced to 35 per cent, from 170 per cent at the beginning of the reform process. This does not mean that there was a smooth and predictable trade policy reform process; on the contrary, there were many policy reversals, as detailed below. Thus, in spite of a gradual movement towards a more open policy stance, the firms during the period we consider in this study still had to regard the policy as highly unstable. This of course reduced the possible positive impact of the reforms.

Still, the trade reforms had a very significant effect on relative prices in the economy, and it was particularly the previously highly protected manufacturing sector which saw its relative prices fall. This we can see in Table 2.1, if we compare the two last columns. The second last column shows that the share of manufacturing in 1995 at 1982 prices (i.e., before the liberalisation process started) was 13.8 per cent of GDP, while at current 1995 prices it was only 9.5 per cent of GDP. This shows that there had been a dramatic relative price shift against manufacturing during the period of reform, while agriculture (for example) saw a price shift in its favour.

2.5.2 The elimination of price controls

The system of price controls that existed in manufacturing reduced incentives to control costs, since the basis for price setting was costs. Exports were further discouraged by the price control measures, since the authorities might demand a lowering of domestic prices if a firm was exporting at lower prices than it set domestically. The government gradually came to the view that price controls were holding investments back and finally abolished them in 1994. The price control act was instead replaced by a Restrictive Trade Practices, Monopolies and Price Control Act aimed at controlling monopolies and collusive tendencies.

2.5.3 Parastatal enterprise reforms

The government had over the years established a whole range of parastatal enterprises. In 1990 there were 135 government majority-owned parastatals in Kenya, while in another 120 the government had minority ownership. This made a total of 255 firms. The share of their output in GDP was 11.6 per cent in 1990. The largest single activity was manufacturing. During 1986–90 the parastatal sector accounted for 16 per cent of gross fixed capital formation. Ownership form has important implications for firm performance.

Table 2.2 Distribution of goods by tariff band, 1987–96 (per cent)

Tariff rates	*1987/88*	*1988/89*	*1989/90*	*1990/91*	*1991/92*	*1992/93*	*1993/94*	*1994/95*	*1995/66*
0	6.9	7.0	5.8	6.1	3.7	2.9	3.1	3.2	3.3
1–10	0.3	0.9	1.6	1.6	4.0	4.6	5.2	4.9	1.8
11–30	30.7	29.6	37.6	37.4	47.6	47.6	56.5	67.8	71.8
31–50	45.4	43.7	23.8	21.6	17.7	20.9	35.2	24.1	23.1
51–60	3.9	5.6	6.0	6.3	3.0	24.0			
61–70	3.8	4.1			24.0				
71–	9.0	9.1	25.2	27.0					
Total	100	100	100	100	100	100	100	100	100

Source: Ministry of Finance.

Table 2.3 Average tariff rates, 1987–96 (per cent)

	1987/88	*1988/89*	*1989/90*	*1990/91*	*1991/92*	*1992/93*	*1993/94*	*1994/95*
Average tariffs:								
Excluding free	42.82	44.38	49.17	47.17	39.39	36.56	40.48	27.27
Including free	39.86	41.29	46.29	44.29	37.92	35.49	39.24	26.39
Average weighted tariffs:								
Excluding free	NA	NA	NA	31.78	27.17	27.25	30.9	22.21
Including free	NA	NA	NA	25.60	23.52	23.20	26.30	18.87

NA = Not available.
Source: Ministry of Finance.

Since the parastatals were always bailed out by the government, they were bound to be inefficient. It was estimated in the 1994/95 Budget Speech that direct and indirect subsides amounted to as much as 5.5 per cent of GDP. The parastatal sector has thus been a drain on the government finances and a drag on economic growth. Kenya is now in a process of privatisation of the parastatals to improve their efficiency and to reduce the budgetary costs.

2.5.4 Export promotion measures

Once the government began seriously thinking about export promotion, it introduced a range of different reforms. Export Processing Zones (EPZs) were set up where firms could produce for export with good facilities and enjoying several tax incentives. These included import duty and value-added tax (VAT) exemptions on imported equipment and raw materials, a 10-year tax holiday followed by a rate of 25 per cent for the next 10 years, exemptions from withholding taxes on dividends and payments to non-residents for the first 10 years and no foreign exchange controls (Gerdin, 1997). From 1994 the EPZ firms were also allowed to sell in the domestic market upon paying import duties, VAT and a low duty. By 1997 no fewer than 15 zones had been established, but still employment in the firm covered was less than 3000.

The Manufacturing Under Bond scheme aims to make it easier for exporters to obtain imported inputs and sell to the export market. This scheme allows also imports free of duty and local purchases free of sales tax, and Green Channel procedures in import licensing. The latter procedures were set up in the late 1980s to reduce delays in obtaining foreign exchange for imports, but became redundant when the foreign exchange market was liberalised.

The Export Compensation Scheme was established in the mid-1970s, but suffered from delays in processing claims and unclear eligibility criteria. The scheme was underfunded and was also open to abuse. It did not, therefore, work well. In the late 1980s the government tried to improve processing of claims. Still, the scheme did not function well and was finally phased out in September 1993.

A Duty/VAT Exemption Scheme was introduced in 1991 which allowed exporters to avoid paying duty and VAT on imported inputs used for the production of goods for a specified export order. In the 1993/94 budget the Essential Goods Production Support Programme was introduced to allow producers of duty-free goods to import raw materials free of duty and VAT. It was administered by the Export Promotion Programme Office (EPPO). This and other similar programmes and bodies set up to administer the export support schemes seem to have suffered from administrative problems, and to have been open to corruption and abuse.

There was also a series of attempts to increase the scope for regional trade and collaboration. In 1983 Kenya entered the Preferential Trade Area (PTA), which covered countries in southern and eastern Africa, and in 1993 the Common Market for Eastern and Southern Africa (COMESA) was established. Then in 1996 a *Secretariat of East African Co-operation* was set up in an attempt to revive the East African Community. At the time of going to press, consultations for ratification of the treaty were under way. So far the impact on actual trade flows among the countries in the region has been limited, but it seems realistic to assume that there will be a substantial expansion of intra-regional trade in coming years, given realistic exchange rates and a much more liberal trading environment.

2.5.5 Investment promotion

Fixed investment in Kenya has, since independence, averaged around 20 per cent of GDP. The capital–output ratio has been around 4, so given current economic structures this investment could generate a 5 per cent growth rate.[5] Though the country has depended heavily on foreign funding, foreign direct investment averaged only US$30 million per year during the 1980s. Foreign investment has been held back by restrictions on dividend transfers as well as by the lack of a stable macroeconomic framework and a credible government. There have been some attempts at simplifying the investment process. In 1982, the government set up an Investment Promotion Centre as a department within the Ministry of Finance. This was later converted into an autonomous parastatal, acting as a one-stop facility for investors and an advisory agency to the government. It is supposed to help the investors through the labyrinth of Kenya's bureaucracy. A foreign investor would need approvals from a number of government departments, including the Central Bank. On top of this, work permits, trade and import licences are also required. As part of the requirement for regional harmonisation, Kenya is making efforts to devise an investment code to guide investments in the country.

2.5.6 Informal sector focus

Informal sector policies changed from harassment in the 1970s to enthusiastic support in the 1980s (Ikiara, 1991), although harassment from local authorities continues. There has been much emphasis on *Jua Kali* enterprises from the launching of the 1989–93 Development Plan onwards. The policy is oriented towards creating an enabling environment via tax and expenditure programmes and better financing. This includes easier credit terms and conditions, via the Kenya Commercial Bank, and simplification of licensing procedures. Much of the support takes the form of formalisation, however, and 'formalization and direct government support would ruin the sector's advantage, e.g. its use of local raw materials, mobility and access to markets, and reliance on self-generated resources' (Ikiara, 1991, p. 318). There are as yet few signs of a significant transfer of informal firms into the formal sector environment.

2.6 Break with the donors and policy reversals

We saw in the previous section that an extensive reform programme was started, and that there was some movement towards a market economy. However, in the early 1990s the donors were unhappy about the slow speed of economic reform, the prevalence of corruption and the slow process of democratisation. They therefore withdrew quick-disbursing aid in November 1991. In the period immediately after this break with the donors there were large swings in policy with regard to foreign exchange and import licensing. The freeze on quick-disbursing aid reduced the supply of foreign exchange, but many donors in actual fact shifted some of this money to project aid instead. The reduction in the inflow of foreign resources was thus less than initially feared. Still, the government needed to find novel ways of attracting foreign exchange. In October of 1991 it introduced Foreign Exchange Certificates (Forex-Cs), which became a major source of foreign exchange for the private sector. This was the first major move in the liberalisation of the currency market, and it attracted back some foreign money.[6] This instrument survived for about two years.

In August 1992, the government announced the establishment of a Foreign Exchange Retention Scheme for exporters of non-traditional goods, such as horticulture, manufactured goods and handicrafts. Exporters would have free access to 100 per cent of their export proceeds, which they could hold in accounts with authorised banks resident in Kenya.[7] A holder of a retention account could utilise the funds to finance business-related payments, including imports of goods and related services, business travel, advertising and marketing expenses, international debt service and remittance of post-tax dividends of foreign owned enterprises. There was no time limit on the use of the funds in the accounts and neither were there any requirements to sell the funds to commercial banks or to the Central Bank. The holders could sell their foreign exchange to the public through the foreign exchange bearer's certificates, at freely negotiated rates. This was an improvement from previous practice, where exporters had to remit their foreign exchange earnings to the Central Bank and apply for foreign exchange for their import needs. As a result of the liberalisation, the free market exchange rate depreciated. Due to the reforms' lack of credibility, many firms held on to their foreign exchange for speculative purposes, which accelerated the depreciation.

After the first multiparty elections at the end of 1992 the government hoped that the relations with the donors would resume. The day before an IMF/International Development Association mission arrived in Nairobi (in February 1993) the government announced sweeping reforms. The shilling (Ksh) was floated, retention accounts were introduced for traditional exports, and the private sector was given a larger role in the marketing of coffee and tea. However, the government had not managed to tighten monetary control, and no agreement was reached. The government then again backtracked on the reforms. The flotation of the shilling was abolished and the retention accounts were scrapped. It was announced that all balances held in these accounts were to be sold to the Central Bank within 48 hours, at the official rate. From then on all export proceeds were to be sold directly to the Central Bank at the official rate. The Bank would then sell foreign exchange to exporters, and the previous system of allocation of licences came back into operation. Import procedures would also return to the old system whereby the Import Management Committee approved import licences and then passed them on to the Foreign Exchange Allocation Committee for the allocation of a Foreign Exchange Allocation License number and subsequent allocation of foreign exchange. This had already prove to be a tedious process, involving a lot of corruption.

These changes were very much opposed by the manufacturers for a number of reasons. First, they lost money when they sold their foreign exchange at the official rate. The market rate had been as high as Ksh 80 to the dollar, while the official rate was only Ksh 45. When the retention accounts were closed, many manufacturers had import bills pending but could not now get foreign exchange from the Central Bank to pay. They were instead advised to finance these imports via the interbank market and at the market rate of exchange. Many importers could not honour their import bills and thereby lost credibility with their foreign suppliers. The latter in turn began to demand cash for their supplies. As a result of these policy changes, the government's export promotion efforts suffered a major setback.

There was strong criticism of this reversal of policies from all quarters. At new meetings with the IMF/International Development Association in April of the second year, the government agreed to resume the reform process. The World Bank in turn agreed to release an Export Development Credit to help stabilise the foreign exchange situation, while the government agreed to try and mop up the excess liquidity, to scrap the much abused pre-shipment export-finance scheme and to clamp down on insolvent or corrupt financial institutions. In May 1993 the Retention Accounts were re-introduced, but with new conditions. First, all account holders would be required to utilise or sell their foreign exchange in the interbank market at market rates within 90 days of crediting their accounts. On expiry of the 90-day period, the funds would be sold to the Central Bank at the official rates. Second, all retention account holders were to retain 50 per cent of their export proceeds and surrender the balance to the Central Bank.

Important changes in the import procedures were announced in May 1993. Importers would now not need (except for a few items) prior approval from the Ministry of Commerce or the Central Bank before undertaking to import goods. Previously, importers went through a lengthy process of import licensing and foreign exchange processing. Now importers were only required to complete an Import Declaration Form, obtainable from commercial banks, and submit it to the same bank for import financing, either from the owner's retention account or from funds acquired in the interbank market. This change was much welcomed by the business community, since it would speed up and simplify importation.

In October 1993 the official and the interbank exchange rates were harmonised, and the rate stabilised at around Ksh 70 per US dollar. Since then the foreign exchange market has been fully liberalised for commercial transactions.

With regard to customs, several changes were introduced to tighten up handling of goods and to make it difficult for goods to enter the market without payment of customs duty and VAT. The Minister of Finance was given power to increase or decrease import and export duty by up to 30 per cent whenever necessary. For some goods, however, including maize, wheat, sugar, milk, rice, and used clothes, the rate of duty could be increased by up to 75 per cent of the normal rate to counteract dumping or other unfair competition from abroad. In addition, duty exemption on such imports as food and used clothes donated to approved non-profit organisations or institutions were removed, in the belief that these types of inflow also represented unfair competition against local producers. Generally, tariff rates have been reduced somewhat, including those on several types of capital equipment, and the number of tariff bands is also smaller than before. A number of tariff exemptions remained available upon application to the Permanent Secretary of the Treasury. The major ones from our perspective are for capital goods aimed at foreign-exchange savings and earnings, and capital goods for investment in small scale industries.

The government also announced the privatisation of reconciliation and export-inspection work required under the EPPO duty/VAT remission programme for imported inputs used in production of exports, for supply to the armed forces, in aid-funded projects, as essential goods with duty/VAT remission, and under the manufacturing-under-bond programme.

The Commissioner of VAT has been given extended powers to collect from persons or firms owing money. Firms selling goods worth at least Ksh 2 million are now taxable under VAT regulations, which have been changed so that there will only be remission of VAT on capital goods worth at least Ksh 5 million per investment, and only if such investment is aimed at foreign-exchange savings or earnings.

Various income-tax reforms were introduced, of which those important for firms are, for example, that importers of capital goods worth more than US$5 million are now allowed to credit import duty paid against income tax payable, provided the importers can prove that the project would generate net economic benefits for Kenya. Manufacturing Under Bond enterprises were allowed an investment deduction for capital expenditure incurred on construction of buildings and on purchase and installation of machinery, including machinery investments as far back as January 1988.

A drought levy was introduced in 1994, but was cancelled in 1995; the corporation tax therefore reverted to 35 per cent. The Kenya Revenue Authority has been established to administer and enforce laws concerning assessment, collection and accounting of all government revenues, in the hope that the system will become more efficient.

The first half of the 1990s, which is the period focused on in this study, was thus a very turbulent period. A range of structural reforms aiming to liberalise the economy were undertaken, but there were many policy reversals. The government had very frosty relations with the international financial institutions and other donors, and Kenyan firms did not view the reform process as stable.

2.7 Growth and investment in the 1980s and 1990s

So what impact did the reform process have on the economy in general and on the manufacturing sector in particular? The latter part of the 1980s saw a fairly stable economic development with increasing per capita incomes (Table 2.4). GDP was growing at about 5 per cent per year and manufacturing about one percentage point more than that. It seemed as if the shift in policy in a somewhat more liberal direction was beginning to pay off but, as we have noted, the whole process was very unstable and the progress on economic reforms soon slowed down.

Kenya entered a serious economic crisis after the withdrawal of donor support in 1991. In the run-up to multiparty elections, held in December 1992, the government furthermore pursued a highly irresponsible economic policy, implying extensive money creation. Political turmoil and ethnic clashes before and after the elections also had serious economic repercussions. Uncertainty about government policies, as well as a shortage of foreign exchange, led to a strong decline in manufacturing investment (Table 2.5). Many investors were not convinced that the government reforms were to be sustained, and adopted a wait-and-see attitude. In 1992 and 1993 GDP grew by just 0.5 per cent and 0.1 per cent, while the manufacturing sector grew by only 1.2 per cent and 1.8 per cent respectively, way below its growth trend. Manufacturing capital formation picked up somewhat in 1993, but was still about 30 per cent below the level reached in 1990.

With the mending of fences with the donors at the Paris meeting of November 1993 and the subsequent stabilisation of the economy, investment activity recovered somewhat. Still, in terms of large-scale investments in manufacturing there was a fairly muted response. Most of the new investment in manufacturing originated from small and new investors, often Asians, including businessmen based abroad. Competition increased in some industries, such as edible oils, which previously had been monopolies.

In 1994 manufacturing showed weak signs of recovery. The poor growth of manufacturing partly reflects the movement of producers into less risky

Table 2.4 Sectoral and overall growth rates, 1986–95

	1986	*1987*	*1988*	*1989*	*1990*	*1991*	*1992*	*1993*	*1994*	*1995*
Agriculture	3.3	4.2	4.6	4.1	3.5	−0.7	−3.3	−3.3	3.0	4.9
Manufacturing	5.8	5.8	6.1	5.9	5.2	3.9	1.0	1.8	1.9	3.9
Building & construction	0.6	4.3	3.9	6.8	3.5	−0.2	−2.3	−3.4	3.3	4.6
Trade, restaurants & hotels	9.8	5.8	5.8	4.4	2.3	1.3	1.5	0.1	6.1	7.9
Transport, storage & comm.	4.3	4.4	4.1	3.0	3.6	3.7	1.7	0.8	3.0	4.3
Finance, real estate & business	6.8	5.2	6.1	7.5	6.4	6.1	6.9	7.5	5.8	6.9
Other	5.1	5.3	4.8	5.5	6.2	4.1	2.7	1.1	1.4	6.2
Government services	6.3	4.8	5.8	5.5	4.4	3.6	2.4	2.1	1.1	1.8
GDP	5.0	4.9	5.1	5.1	4.2	2.1	0.5	0.3	3.0	4.9

Source: *Economic Survey*, various issues.

Table 2.5 Annual growth rates of capital formation, 1986–95

	1986	*1987*	*1988*	*1989*	*1990*	*1991*	*1992*	*1993*	*1994*	*1995*
Agriculture	5.4	9.3	1.8	−8.6	1.3	−6.5	5.9	−9.8	28.0	30.1
Manufacturing	28.1	10.1	22.2	−2.4	18.7	−14.8	−12.6	5.7	16.0	41.0
Building & construction	15.4	28.0	10.5	19.2	20.3	13.3	−12.4	−10.6	−31.7	17.9
Transport, storage & comm.	35.9	4.3	−23.4	1.3	−3.1	21.8	2.6	38.4	38.4	2.4
Government services	4.5	−3.1	44.1	−14.4	9.9	−23.6	14.0	−42.5	37.8	26.3
Trade, restaurants & hotels	−41.2	−6.1	45.3	−40.8	27.4	222.5	−20.0	20.2	−46.3	16.7
Finance, real estate & business	−41.9	56.5	95.4	9.9	7.0	−2.4	−22.9	2.0	47.8	25.1
Other	12.8	5.4	−9.0	16.6	−23.8	−20.9	−11.4	−4.6	2.2	−4.7
Total	13.1	6.7	9.3	−1.0	4.3	−1.9	−4.0	−2.8	14.6	16.9

Source: *Economic Survey*, various issues.

activities, such as retail trade and wholesale. A major observation from our own interviews was that the entrepreneurs were in general very uncertain about the direction of policies and politics in Kenya. It was therefore natural to take a wait-and-see attitude to investments, and to try to diversify into less risky activities. The government has changed policies for the better according to most of the producers we interviewed, but there were major complaints about the administration of the new policies: corruption and rent-seeking activities seemed to be increasing. This general feeling of mistrust was holding back private investment. During the second half of the 1980s investment was close to 15 per cent of GDP, but then it started to decline, reaching its lowest level in 1992, 9.7 per cent (*Economic Survey*, 1995). In 1993 it increased to 10.4 per cent, while in 1994 it again declined slightly to 10.2 per cent. Net long-term private flows, including foreign investments, were also negative for the third year in a row, indicating that foreign investors had not regained confidence in the Kenyan government. Thus the structural adjustment experience of Kenya was similar to that of many other African countries, in that private investments was very slow to respond to changes in the structure of incentives. However, there was some recovery of public investments to 10.4 per cent of GDP in 1994 from a low of 7.8 per cent in 1992. There was an increase in infrastructure investment during the year, mainly in energy and roads, but many years of high investments will be required to repair the run-down system.

During the 1993–95 period the government had to devote much effort to short-term crisis management to deal with the monetary overhang from the election spree. The lack of fiscal adjustment delayed the process and reduced the incentives for exporters (Levin, 1998). In the middle of 1994 there were some signs of a fragile recovery, threatened by political instability. GDP grew by 3 per cent in 1994, which was a considerable recovery from the year before. However, growth was mainly concentrated in the agricultural and service sectors, while manufacturing growth was only 1.9 per cent, and the share of manufacturing in GDP has remained virtually constant over the last 10 years. Thus, in spite of the declared aim of creating an environment conducive to manufacturing growth, the results were still poor. Various manufacturing industries had very different growth rates, as we see in Table 2.6, which shows the development of manufacturing sub-sectors between the mid-1980s and mid-1990s. With regard to the four sectors that we specifically analyse in this book we may note that textiles and clothing were shrinking dramatically during the years of our surveys. Food, wood and furniture were less dramatically affected, while the metal sector did fairly well.

2.8 Long-term productivity growth in manufacturing

The growth of manufacturing production in Kenya has been driven by increases in factor input, not by improvements in productivity. As a matter

Table 2.6 Annual changes in manufacturing output by industry (per cent)

	1986–90	*1991*	*1992*	*1993*	*1994*	*1995*
Meat and dairy products	11.1	−10.6	−24.9	1.9	−9.2	29.1
Canned vegetables, fish, oils, and fats	10.2	13.2	−6.7	13.4	−28.5	6
Grain mills products	−1.9	0.1	−4.3	−9.2	32.3	18.6
Bakery products	4.4	5.1	4.1	3.2	53.8	9
Sugar and confectionery	3.2	3.7	−7.1	1.3	20.9	25.4
Miscellaneous foods	8.6	0.3	4.1	−7.1	2.7	2.1
Beverages	7.7	−3.3	16.2	−0.7	−7.5	4.7
Tobacco	3.5	−2.6	11.1	1	0.7	8.4
Textiles	5.5	−4.1	0	15.3	−25.8	−27
Clothing	−0.5	−6.8	−0.9	−8.8	−37.1	−16.8
Leather and footwear	5.3	2.3	−4.3	−9.5	10.1	−32.1
Wood and cork products	1.2	4.1	1.2	−0.4	3.1	−3.4
Furniture and fixtures	0.3	−3.9	−33.9	6	2	3.8
Paper and paper products	7.4	5.2	20.1	−30	−13.8	−1.7
Printing and publishing	3.7	1	1.3	0	3.3	6.3
Basic industrial chemicals	6.7	10.6	0.1	4.5	−12.9	−1.9
Petroleum and other chemicals	16	11.6	−5.7	−2.6	−3.3	8.5
Rubber products	6.1	−1.1	95.5	−1.3	−1.4	1.2
Plastic products	2.6	20.6	18	10.3	1.5	6.1
Clay and glass products	6.7	−2	59.9	108.4	47.6	18.6
Non-metallic mineral products	5.9	4.4	17.9	−2	5.1	−1
Metallic products	17.4	14.7	−9.5	0.2	11.3	0.9
Non-electrical machinery	−3.1	−2.7	−5.6	−0.7	6	−22.1
Electrical equipment	4.2	36.5	−3.2	−10.7	0.8	12.2
Transport equipment	3.4	−1.7	−8.8	−10.4	5.6	−7.4
Miscellaneous manufactures	14	8.6	0.4	0	0	6.6
Total manufacturing	6.2	2.8	1.3	1.8	1.9	3.7

Source: *Economic Survey*, various issues.

of fact, productivity improvement has been non-existent. Gerdin (1997, ch. 6), in a detailed analysis of the growth of total factor productivity (TFP) growth in Kenyan manufacturing between 1964 and 1994,[8] shows that the mean TFP growth was −0.12 per cent. Up until the 1970s it appears that change in scale was the most important source to TFP growth, while technical change was more important than scale changes thereafter. Given the inward orientation of the sector and the small size of the domestic market, there was little scope for exploitation of scale economies in Kenya. Technical change was positive only in 1975, 1980, 1983 and 1989–92. TFP growth, however, was negative in 1992–94. Any technical progress around 1990s is likely to be due to the gradual liberalisation of the economy.

TFP in manufacturing thus actually declined in all the three RPED years. Causes that might account for this are shortages of foreign exchange and thus problems in getting access to imported raw materials and higher input prices. The devaluation in early 1993 made foreign exchange very expansive and inflation shot up to 46 per cent. The government tried to contain domestic money supply by issuing Treasury Bills, which depressed domestic demand. There was also extensive political turbulence in 1992 in the run-up to the first multiparty elections. Still, there seem to have been some improvements in the later years as the economy has been undergoing structural adjustment measures and is being gradually liberalised. Generally one would expect the gradually increasing skill-composition of the labour force to affect TFP positively, while the slow rate of investment would tend to hold it back.

Still, the overall conclusion is that TFP growth or technical progress has been very low and actually not significantly different from zero during the post-independence industrialisation drive. Once the easy import-substitution phase came to an end, there was not enough scope for productivity growth in the economy to make it possible to break out of the domestic market and into the global marketplace on any significant scale. From about the mid-1970s Kenya thus combined a declining rate of investment with TFP stagnation or decline. This shows that the environment for a manufacturing breakthrough was not in place. Possibly this could change with the turnaround of policy that was undertaken in the 1990s.

2.9 Concluding remarks

Kenya experienced relatively high growth rates up until the first oil crisis in 1973. During the 1960s and early 1970s there was an investment boom in Kenya. Between 1964 and 1971 gross fixed capital formation grew by 17.2 per cent per year in constant prices, while the growth rate was only 0.1 per cent per year between 1971 and 1993. Between 1994 and 1996, however, there was a substantial recovery, but this was broken again by several years of stagnation.

There was a rapid build-up of manufacturing capital in the 1960s, much of it in relatively recent machinery and equipment (Ritter, 1988). Eventually excess capacity was built up, and since several of the industries were meant to serve the wider East African market they suffered when the community collapsed. Given the excess capacity there was less need for new investment. The investment boom of the 1960s was largely built on MNCs, but given that the wider East African market was no longer available from the Kenyan base and given that Kenya was not perceived as a location from which MNCs could penetrate the world market, MNC investment slowed down dramatically in the mid-1970s. The global recession in the early 1980s, plus the coup attempt in 1982, further reduced incentives to invest. Several MNCs actually withdrew or sold off their investments in Kenya in the late 1980s. During the coup attempt there was looting of Asian property, which further increased their sense of insecurity. They did not feel that the government was willing to guarantee their property.

Why, more specifically, has the growth of the Kenyan manufacturing sector been so poor? First, in Kenya (as in most other African economies) there is a high level of economic and political risk. The economy is vulnerable to external shocks, and the weakness of the financial system makes it hard to hedge against those risks. High risks mean that the cost of capital is bound to be high, and this implies that investments will be low. Second, poor infrastructure has reduced returns on prospective investment projects. Extensive power rationing in 2000 is but one indication that little has been achieved in terms of infrastructure development. Third, the government set up a control regime with extensive trade restrictions and other barriers to private sector investment. Inward-oriented policies are particularly costly in small economies such as the Kenyan one. Fourth, the firms themselves often argue that there is a serious credit constraint. This is true in some cases, especially for small producers, but credit is often available for good projects or sound inventors. Moreover, the financial constraint can be solved, at least in the medium term, from firms own' savings or retained profits. Fifth, the lack of demand for products is a severe investment constraint. Investors do not invest unless there is a market for output, and if the domestic market is not growing and the firms are uncompetitive in the foreign market, then there will be little investment. The low levels of capacity utilisation observed show that firms often are able to meet new demand from existing capacity.

Africa's export is heavily concentrated in unprocessed primary products, indicating that this is where Africa has its comparative advantage. Wood (1994) suggests that this is natural, since Africa has an abundance of labour and natural resources, while it is short of physical as well as human capital. Manufacturing production, on the other hand, requires much human capital and few natural resources. Wood argues that it is likely that Africa's trade pattern will over time come to resemble that of Latin America rather than that of Asia, which is a situation where the abundance of natural resources continues to explain the pattern of trade.[9] However, Kenya has managed to take some steps towards manufacturing exportable production and now manufactured exports (broadly defined) constitute more than 20 per cent of total exports.

To understand manufacturing growth we must understand productivity change. It is worth asking whether the reforms and inducements introduced so far are sufficient to make Kenya competitive beyond agro-processing, horticulture and tourism. A supportive environment that is sufficiently efficient and attractive to investors is not yet in place, although there have been some improvements in the recent decade. Kenya does need both higher levels of manufacturing investment and productivity improvements to achieve higher growth in the sector. So far the response to the regime shift has been limited. The reforms that are needed are straightforward, but the political system is not able to deliver and maintain an environment that is competitive in a globalised world. Investments in Kenya are exposed to a high political risk, and this means that the highly mobile international investment capital still goes elsewhere.

Notes

1. One of the most important reports on the transition to independence was that of the East African Royal Commission 1953–55 (1955). The Commission warned against protectionist policies, and voiced the fear that once they were in place they would be hard to remove. Instead, they argued, the government should concentrate on providing a enabling environment (although they did not use that term) for business.
2. An alternative view is due to Coughlin (1988, p. 276), who argues that 'the nation must concentrate on deepening industrialization and begin to manufacture many of the components, raw materials, and machinery used by industry and agriculture'. He argues for further import substitution and that import substitution has just begun. He claims that there has been too little infant industry protection or that it was abandoned too early and that this has wiped out some industries (p. 277).
3. See, e.g., McCormick (1991), who discusses the implication for development of the emergence of the informal sector. She concludes that small firms mainly grow by proliferation. This casts doubt on whether they can be the engine of growth and technological improvement.
4. This is clearly shown in the 1989–93 *Development Plan* (Kenya, 1989), where it is stated that the emphasis from now on is to be on the promotion of export of consumer and intermediate goods. The reform measures envisaged had been outlined already in the Sessional Paper No. 1 of 1986, entitled 'On Economic Management for Renewed Growth'.
5. Damus (1990) argues that the low estimates of total factor productivity growth presented in some studies of Kenya can, to a certain degree, be explained by the fact that low productivity sectors have increased in weight over time. This, he argues, is in turn due to market distortions. His crude estimates suggest that due to price distortions there was a waste of up to 2.5 per cent of the labour input. Beaulieu (1990) notes that Kenya's competitive edge in the export markets has been reduced by import substitution on the input side. Domestic input costs have increased relatively faster than if inputs had been imported.
6. According to International Financial Statistics foreign deposits held by Kenyans abroad amounted to US$2.17 billion in September 1993.
7. Traditional products, e.g., meat products, coffee, tea, wheat and maize products, pyrethrum products, wattle extracts, hides and skins, cotton, sisal and metal products were not eligible
8. He applies a model that allows for temporal variations in technical change and uses panel data techniques for estimation on the basis of data from nine manufacturing sub-sectors.
9. Mauritius has a resource endowment similar to that of East Asia and it also has a trade pattern similar to the East Asian one: that is, a dominance of manufactures.

References

Beaulieu, E. (1990), *Structural Change in Kenya 1967–86*, Technical Paper No. 90–09, Long Range Planning Unit, Ministry of Planning and National Development, Nairobi.

Bevan, D, Collier, P. and Gunning, J., with Bigsten, A. and Horsnell, P. (1990), *Controlled Open Economies*, Oxford University Press, Oxford.

Bigsten, A. and Moene, K.-O. (1996), 'Growth and Rent Dissipation: The Case of Kenya', *Journal of African Economies* 5(2).

Coughlin, P. (1988), 'Towards a New Industrialization Strategy in Kenya', in Coughlin and Ikiara (1988).

Coughlin, P. (1990), 'Moving to the Next Phase', in Riddell (1990).

Coughlin, P. and Ikiara, G.K. (1988), *Industrialization in Kenya. In Search of a Strategy*, Heinemann, Nairobi and James Currey, London.

Coughlin, P. and Ikiara, G.K. (eds) (1991), *Kenya's Industrialization Dilemma*, Heinemann, Nairobi.

Damus, S. (1990), *Productivity Change in Kenya 1971–1986*, Technical paper No. 90-08, Long Range Planning Unit, Ministry of Planning and National Development, Nairobi.

East African Royal Commission 1953–1955 Report (1955), London.

Eglin, R. (1978), 'The Oligopolistic Structure and Competitive Characteristics of Direct Foreign Investment in Kenya's Manufacturing Sector', in Kaplinsky (1978).

Gerdin, A. (1997), *On Productivity and Growth in Kenya*, Ekonomiska Studier utgivna av Nationalekonomiska institutionen, Handelshögskolan vid Göteborgs Universitet, No. 72, Göteborg.

Ikiara, G.K. (1991), 'Policy Changes and the Informal Sector: A Review', in Coughlin and Ikiara (1991).

ILO (1972) Kenya: Employment, Incomes, and Equality, ILO, Geneva.

Juma, C. (1991), 'Investment Strategy and Industrialization Policy: Lessons from Power-Alcohol Development', in Coughlin and Ikiara (1991).

Kaplinsky, R. (ed.) (1978), *Readings on the Multinational Corporation in Kenya*, Oxford University Press, Nairobi.

Kenya, *Budget Speeches*, Nairobi (various issues).

Kenya, *Economic Survey*, Nairobi (various issues).

Kenya (1965), *Sessional Paper no. 10 on African Socialism and its Application to Planning in Kenya*, Nairobi.

Kenya (1989), *Development Plan 1989–93*, Nairobi.

King, K. and Aboudha, C. (1991), *The Building of an Industrial Society. Change and Development in Kenya's Industrial Sector 1970–1990*, Occasional Paper No. 30, Centre of African Studies, Edinburgh University, Edinburgh.

Levin, J. (1998), *Structural Adjustment and Poverty: The Case of Kenya*, Ekonomiska Studier utgivna av Nationalekonomiska institutionen, Handelshögskolan vid Göeborgs Universitet, No. 84, Göteborg.

McCormick, D. (1991), 'Success in Urban Small-scale Manufacturing: Implications for Economic Development', in Coughlin and Ikiara (1991).

Nyongo, A.P. (1991), 'The Possibilities and Historical Limitations of Import-Substitution Industrialization in Kenya', in Coughlin and Ikiara (1991).

Riddell, R.C. (ed.) (1990), *Manufacturing Africa: Performance and Prospects of Seven Countries in Sub-Saharan Africa*, James Currey, London, and Heinemann, Portsmouth.

Ritter, A.R.M. (1988), 'Productivity Change in the Non-Agricultural Economy in Kenya, 1964–1987', Technical Paper No. 88-06, Long Range Planning Unit, Ministry of Planning and National Development, Nairobi.

Sharpley, J. and Lewis, S. (1990), 'The Manufacturing Sector in the mid-1980s', in Riddell (1990).

Swainson, N. (1980), *The Development of Corporate Capitalism in Kenya 1918–1977*, Heinemann, London.

Vaitsos, C. (1991), 'The State and Foreign Business Interest', in Coughlin and Ikiara (1991).

Van Arkadie, B. (1964), 'Import Substitution and Export Promotion as Aids to Industrialization in East Africa', *East African Economic Review* 1.

Wood, A. (1994), North-South Trade, Employment and Inequality. Changing Fortunes in a Shill-Driven World, Clarendon Press, Oxford.

3

The Regional Programme of Enterprise Development Survey, 1993–95

Renato Aguilar and Arne Bigsten

3.1 Introduction

This survey of Kenyan manufacturing firms was undertaken three times. The first round of interviews was done in February–March 1993, the second in May–June 1994 and the third in August–September 1995. Much of the data gathered refers to the most recent financial year of the firms, which means that such data refers to the period 1992–94. When firms were asked about their current situation, their replies refer to the actual time of the interview. The towns covered were the four major towns of Nairobi, Mombasa, Nakuru and Eldoret. Our survey did not cover the whole of manufacturing, but concentrated on four sub-sectors: food processing, textiles and garments, wood working, and metal working. However, since these four sub-sectors comprise about 73 per cent of manufacturing employment, we still provide a reasonably comprehensive picture of the manufacturing sector in Kenya. In terms of employment the food sector is the largest of the four. It provides 32 per cent of the employment in our four sub-sectors, compared to 26 per cent for textiles and garments, 25 per cent for metal, and 17 per cent for wood. If we instead look at employment by firm size, we see that 20 per cent is in the category 1–5 employees, 17 per cent in 6–20, 7 per cent in 21–75, 19 per cent in 76–500, while as much as 36 per cent is in the 500+ category.

The joint team of the Economics Departments of Göteborg and Nairobi which undertook the surveys have presented preliminary analyses of the data and quite detailed tables of results in a series of publications (Departments of Economics, 1993a, 1993b, 1994, 1995a, 1995b, 1996). The analysis in this book extends and refines this work.

3.2 The four sub-sectors

All the four sectors chosen for analysis emerged early in the country's industrialisation process. In 1961 they produced 68.4 per cent of manufacturing value added in Kenya, and the share was still high in 1990, when it was 65.6 per cent. In terms of productivity (value added/employment) the food sector lies considerably above the manufacturing average, while the other three lie well below. The wood-working sector in particular is characterised by a lower productivity, although the gap *vis-à-vis* the rest has decreased over the last three decades. In 1990 the labour productivity index in food processing was 126.7 (manufacturing average = 100), while it was 86.3 in fabricated metal, 56.7 in textiles and garments, and 44.7 in the wood-working sector. The gap among the earlier three sectors was rather stable over the period 1961–90, while the index of wood working moved up from 17.7 in 1961 to 44.7 in 1990.

In terms of wage rates the dispersion among sectors has declined. Formal wood-working wages have approached the levels in the other sectors. With the exception of fabricated metal, which had the highest wages both in 1961 and 1990, the structure of relative wages shows a pattern similar to that for labour productivity. In 1990, the wage index (manufacturing average = 100) was 95.0 in food processing, 76.2 in textiles and garments, 67.9 in wood working and as high as 125.5 in metal working. The last three are not explained by a higher capital–output ratio, which instead is highest in food processing. The capital–output ratio index for the latter was 106.4 in 1990 (manufacturing average = 100), while it was 46.9 for textiles and garments, 59.6 for wood working and 61.0 for metal working. The higher wages are possibly due to the fact that the skill level in metal working is higher than that in the other sectors.

Let us now briefly look at some characteristics of the chosen sectors. In terms of the size distribution of firms, they are characterised by an above-average proportion of small-sized manufacturing activities. Each sector includes a very diverse range of production activities, as highlighted below.

The range of activities in food processing spans grain milling, production of edible oils, sugar processing, dairy processing, animal slaughtering, canning and preservation of meat, fruits and vegetables, bakery and confectionery, salt production, production of beverages, production of food preservatives and seasoning, production of animal feeds, and so on. Aspects of production which are specific to the sector are the seasonality in the production cycle, a preponderance of highly perishable commodities, with implications for the technology for handling, transporting, storage and processing and variability in quantity and quality of raw materials.

The textile sector uses natural and chemical fibre processing (spinning, weaving and knitting) in garment manufacture, manufacture of household items (furnishings, carpets) and the production of industrial goods (belting, upholstery, rope and twine, sacks, etc.). Basically, garment manufacturing is the dominant activity but it is quite fragmented, and is by far the least homogeneous in terms of products.

Wood and furniture industries undertake timber processing followed by the manufacture of furniture, wooden sculptures, storage and packaging materials (chests, crates, racks, shelves) and joinery materials for the construction industry (roofing, flooring, doors and windows, pre-fabs, etc.).

The Kenyan metal sector covers a wide range of products and levels of technological sophistication, ranging from simple and easy-to-manufacture products to more complicated ones. Production units also vary from small shops to large international companies. In terms of products, a majority of Kenyan firms undertake simple engineering work and repairs based mostly on sheet metal (containers, utensils, window frames, metal furniture, etc.). The type of machinery and equipment used is basic, as are the skills required.

It may be important to note that even after going through several years of import substitution, many firms still continue to organise their operations around a limited collection of production equipment and tools. There, some machinery and equipment is manufactured, including hot forged, hot pressed and stamped products and basic machined products (screws, taps, gears, shafts, pulleys, etc.).

3.3 Sampling procedure

The primary sample consisted of a total of 200 firms in manufacturing, 50 each in food processing (ISIC 31), textiles and garments (32), wood working (33), and metal working (38). The sample included both the formal and the informal sectors with 75 per cent of the interviews in formal firms and 25 per cent in informal ones. We restricted the sample to Nairobi, Mombasa, Nakuru and Eldoret, which covers the majority of formal manufacturing. The formal firms were chosen from the Central Bureau of Statistics' master file of registered firms. This is in principle continuously updated, but it did have numerous problems such as the inclusion of firms which either had never started or were defunct. Still, it was the best source available. We defined five size-strata for each of the four industrial sectors included in the study. These strata include firms from 1 to 5 employees, from 6 to 20, from 21 to 75, from 76 to 500, and 501 or more employees. Thus, a total of 20 strata were defined for the formal sector. Firms were randomly chosen within each stratum according to their distribution among the four cities covered by the study. The selection process slightly overrepresented employment in the larger formal firms. Apart from the stratification by size, the selection was itself random across the four cities included in the study. Out of our primary sample of 150 formal firms, 12 firms declined to be interviewed. A few other firms were not found, had closed or switched into other activity. These firms that could not be interviewed were replaced by firms with the same characteristics from our reserve list.

To construct a sample frame for the informal sector we undertook a primary listing of firms engaged in the four sectors in the four cities. The listing for Nakuru and Eldoret was close to complete, while for Nairobi and

Mombasa we only covered some sections of the towns. On the basis of this listing, we estimated the distribution of firms across the four sub-sectors, and found this to be 30 per cent textiles, 28 per cent metal working, 30 per cent wood working and 12 per cent food processing. If we had tried to represent the employment in the informal sector as well as formal employment, we would have had to choose most of our sample from this sector. However, since the degree of variation in the informal sector is much smaller than in the formal one we decided to include a relatively small number of informal firms. The primary sample of 50 firms was then chosen in the given proportions. This means that in textiles 70 per cent of the sample is in the formal sector, while the relevant figures are 72 per cent in metal working, 70 per cent in wood working and 88 per cent in food processing. With regard to the informal sample we assumed that the best predictor of its distribution among the cities was the population. We therefore selected 65 per cent of the informal firms in Nairobi, 27 per cent in Mombasa, 8 per cent in Nakuru and 5 per cent in Eldoret. In the case of the informal sector, no stratification according to size was necessary. Thus, this sub-sample includes only four strata, one for each industrial branch.

With regard to the informal firms, no introductory letter was sent. We went to see them directly and tried to arrange an interview. This worked very well in most cases. Since our listing was done less than six months before the time of the first interview we were almost always able to locate the firms we wanted to interview. However, when this was impossible we chose a similar firm in the same area.

We also added the 19 firms from the pilot survey to the first round sample, although there are a few bits of information missing from the first round interviews since the questionnaire did not change much between that and the main survey. We took this opportunity to beef up the sample a bit with some extra interviews. Thus, in all we interviewed 224 firms the first year, 56 in textiles, 59 in wood working, 59 in metal working, and 50 in food processing. The sample in our four sub-sectors was thus 224 in the first round.

A total of 276 firms were interviewed at least once. We interviewed 224 firms in 1994, 216 firms in 1994, and 218 firms in 1995. Table 3.1 shows the time and frequency of the interviews.

The sample had a significant rate of attrition, although smaller than expected. Table 3.2 shows the attrition rates. Of the 224 firms interviewed in 1993, just 29 firms were lost in the second wave and a further 26 were lost in the 1995 wave. However, we could again interview one firm which was in the 1993 sample but not in the 1994 one. Thus a total of 54 firms were lost between 1993 and 1995. Out of 216 firms interviewed in 1994, we were unable to interview 30 of them in 1995. In a few cases firms declined to be interviewed again, while in other cases we know that they had ceased to operate. Finally, there were instances when we could not

Table 3.1 Frequency of interview of firms

	Firms	*%*
Interviewed only once	63	22.8
In RPED93	28	–
In RPED94	31	–
In RPED95	4	–
Interviewed only twice	44	16.0
In RPED93 and RPED94	26	–
In RPED93 and RPED95	1	–
In RPED94 and RPED95	17	–
Interviewed in all the three waves	169	61.2
Interviewed at least once	276	100.0

Table 3.2 Attrition rates of the surveys

	1993		*1994*		*1995*	
	Firms	*%*	*Firms*	*%*	*Firms*	*%*
1993	224	0.0	29	12.9	54	24.1
1994	–	–	216	0.0	30	13.9
1995	–	–	–	–	218	0

retrace the firm or the entrepreneur and were unable to conclude with certainty whether the firm had actually closed down or just moved. Because of these problems in determining exactly the reason for exit, we have not undertaken any analysis of why firms cease to operate.

Thus, we have a panel of 169 firms in three waves, plus a small panel of 44 firms with only two cross-sections. A total of 63 firms can only be used in cross-section studies with the distribution shown in the table.

The firms that were interviewed were generally very cooperative. They were willing to discuss even sensitive issues, so we do believe that the responses in most cases are representative. However, the informal firms in particular did not keep comprehensive records, which means that we had to rely on their recall. That may make some of the financial variables somewhat approximate, but we do not believe that they systematically tried to provide misleading information.

3.4 Stratification and weighting

Since we have a stratified sample, particularly between formal and informal firms (Table 3.3), we have constructed weights that make it possible to make representative employment-oriented as well as firm-oriented inferences.

Table 3.3 Sample distribution by formal and informal sectors

	Employment	*Firms*
Formal	26 147	163
Informal	249	61
Total	26 396	224

The employment weights, which also may be taken as proxies for output weights, are weights which give each employee in a firm the same probability of being included in the sample. The firm weights are weights giving each firm the same probability to be sampled. We have weights for the whole sample as well as for the formal and informal categories separately. These weights are useful for industry-wide inferences as well as for inferences about the informal and formal sectors respectively.

To compute our weights we needed to know the actual levels of employment in our four sub-sectors, namely textiles, wood working, metal working and food processing. Our estimates refer to 1992. The aggregate level of formal manufacturing employment in 1992 was 189 600 (Kenya, 1993, p. 56), while that of aggregate informal manufacturing employment was 115 116 (p. 63). Disaggregated data for the formal manufacturing sector (data from diskette from the Ministry of Economic Planning with data on the Employment and Earnings Survey, 1991) show that 72.9 per cent of those employed in manufacturing were to be found in our four sub-sectors. We assumed the same fraction for both the formal and the informal sector. These assumptions gave us our final estimate of total employment in 1992 in our four sectors, namely 138 220 in the formal sector and 83 920 in the informal sector or a total of 222 140.

The four sectors in the formal sector list from the Central Bureau of Statistics had 803 firms employing 99 800 people. Since our estimate of formal employment for the four sectors is 138 220, we have estimated that there are 1112 formal firms in the population. Based on the estimates of informal employment and the average size of the informal firms in our sample (4.08 employees), we estimate the total number of informal firms to be 20 569. The total number of firms in the four sectors is thus estimated to be 21 681.

We assumed that the distribution of employment and firms by sector and employment bracket in the Central Bureau of Statistics formal sector list was a good estimation of the same distribution for the population of formal firms. Thus weights for the formal sector were computed by breaking down the list accordingly.

Our own enumeration of firms in the informal sector in the four towns included in the survey gave us a basis for the distribution of informal firms among our four sub-sectors. We found that 30 per cent were in textiles,

28 per cent in metal working, 30 per cent in wood working and 12 per cent in food processing. We broke down the informal sample in these proportions only.

It should be noted that this sample is highly stratified, at least between formal and informal firms. At least two dimensions are important for the definition of these weights. One dimension is the number of firms, and the other one is the size of the firms. In our case we used employment as a proxy for the size of the firms. For example, if we want to say that a given per centage of the firms or the entrepreneurs have a given characteristic, weights based on the number of firms should be used. On the other hand, if we want to say that firms are responsible for a certain share of total output we should use size-oriented weights.

We constructed Tables 3.4 and 3.5 using the published data on industrial employment. However, we had to make a number of assumptions in order to transform aggregate information into sectoral data. To a large extent we based these assumptions on the analysis of the population lists used for the sampling procedure.

Table 3.4 Informal sector structure

	Firms	*Employment*
Textile	6 171	25 176
Wood	6 171	25 176
Fabricated Metal	5 759	23 498
Food	2 468	10 070
Total	20569	83920

Table 3.5 Formal sector structure

Branch	*Size*	*Firms*	*Employment*
Textile	1–5	86	304
	6–20	52	725
	21–75	58	3.199
	76–500	45	10.033
	501 +	13	17 253
	Total	254	31 514
Wood	1–5	26	150
	6–20	26	585
	21–75	30	2 443
	76–500	16	4 845
	501 +	2	4 417
	Total	100	12 440

Table 3.5 continued

Branch	*Size*	*Firms*	*Employment*
Fabricated Metal	1–5	52	233
	6–20	80	1 183
	21–75	74	3 515
	76–500	54	11 747
	501+	2	15 942
	Total	262	32 620
Food	1–5	53	77
	6–20	133	800
	21–75	124	2 528
	76–500	158	15 427
	501+	28	42 816
	Total	496	61 648
Total		1 112	138 222

Different systems of weights (or expansion factors) could be devised based on these data and in accordance with the reference universe underlying our inferences. Thus we constructed weights for the sample of the formal sector, for the sample of the informal sector, and a system of weights for the total sample.

References

Departments of Economics, University of Göteborg and University of Nairobi (1993a), *Economic Development and the Manufacturing Sector in Kenya*, World Bank, Washington, DC and Department of Economics, University of Göteborg.

Departments of Economics, University of Göteborg and University of Nairobi (1993b), *Regional Programme of Enterprise Development. First Report on the Kenyan Survey*, World Bank, Washington, DC.

Departments of Economics, University of Göteborg and University of Nairobi (1994), *Limitations and Rewards in Kenya's Manufacturing Sector: A Study of Enterprise Development*, World Bank, Washington, DC.

Departments of Economics, University of Göteborg and University of Nairobi (1995a), *Regional Programme of Enterprise Development. First Report on Round 2 of the Kenyan Survey*, World Bank, Washington, DC.

Departments of Economics, University of Göteborg and University of Nairobi (1995b), *Manufacturing in Kenya Under Adjustment. Final Report on the Round 2 RPED Survey*, World Bank, Washington, DC.

Departments of Economics, University of Göteborg and University of Nairobi (1996), *Regional Programme of Enterprise Development. Report on Round 3 of the Kenyan Survey*, World Bank, Washington, DC.

Kenya (1993), Economic Survey 1993, Nairobi.

Part II

Structure of Manufacturing

4
Firm, Product Market and Technological Characteristics

Gerrishon K. Ikiara, Peter Kimuyu, Mutsembi Manundu and Wafula Masai

4.1 Introduction

This chapter describes the main characteristics of Kenya's manufacturing sector during the first half of the 1990s. This description seeks to shed some light on the nature of the sector and lay the foundation for more detailed analysis. Unlike in other chapters, we do not use weights but instead present results by enterprise size where possible. That way, we are able to take into account the low positive response rates on some of the key variables used in the chapter variables.

Section 4.2 looks at the structural attributes such as the size distribution of enterprises, historical development of firms and the interplay between ethnicity and legal status, gender, and ownership structure. Section 4.3 focuses on technology and related issues, while section 4.4 examines the structure of the product markets and competition within the sector. Section 4.5 analyses the exposure to and resolution of business disputes before summarising in section 4.6

4.2 Structural characteristics of the sector

4.2.1 Size distribution

By the mid-1990s, a large part of manufacturing employment in Kenya was generated within the small-scale sector, which had grown rapidly as a result of the steady growth of informal activities since the 1980s. In the mid-1990s the small-scale sector accounted for about two-thirds of new job opportunities created in the Kenyan economy (Kenya, 1997). At the start of the RPED survey, micro enterprises (1–5 employees) and small enterprises (6–20) accounted for more than a third of manufacturing employment, and medium-sized enterprises (26–75) less than 10 per cent. Larger firms with more than 76 employees accounted for about 55 per cent of the employment. The data reflect the limited employment potential of medium-sized firms in Kenyan manufacturing, mimicking the 'missing middle' phenomenon that is common in the developing world.

The distribution of the sample by firm size and sector (Table 4.1) reveals that two-thirds of the micro enterprises are textile and metal working, the textile ones accounting for 40 per cent of such enterprises. Many of the small firms are involved in food processing and wood working. The wood sector has the largest share of the medium-sized enterprises, while the food and textile sectors have a greater share of the very large enterprises. Whereas most small, informal firms generally use simple technologies, have low start-up capital, low management requirements and often rely on family labour, these attributes are more common among textile and garment firms.

4.2.2 The enterprise life cycle

The majority of the enterprises were established in the post-independence era, with only 13 per cent, having been established before 1963 (Table 4.2). Nearly half of the firms were established during 1963–80. Most of the smaller enterprises are recent and only a small number of the micro and

Table 4.1 Distribution of the enterprises by size and sector

	Food		*Wood*		*Textiles*		*Metal*		*Total*	
	No.	*%*	*No.*	*%*	*No.*	*%*	*No.*	*%*	*No.*	*%*
Micro	6	10.2	13	22.0	24	40.7	16	27.1	59	100.0
Small	14	31.1	12	26.7	9	20.0	10	22.2	45	100.0
Medium	16	23.5	20	29.5	16	23.5	16	23.5	68	100.0
Large	11	26.2	12	28.6	7	16.6	12	28.6	42	100.0
Very large	3	30.0	2	20.0	3	30.0	2	20.0	10	100.0
Total	50	22.34	59	26.3	59	26.3	56	25.0	224	100.0

Source: Computed from the first wave.

Table 4.2 Distribution of enterprises by size and date of establishment

	Before 1963		*1963–80*		*1980–93*		*Total*	
	No.	*%*	*No.*	*%*	*No.*	*%*	*No.*	*%*
Micro	4	6.8	21	35.6	34	57.6	59	100.0
Small	6	13.3	17	37.8	22	48.9	45	100.0
Medium	11	16.2	41	60.3	16	23.5	68	100.0
Large	2	4.8	29	69.0	11	26.2	42	100.0
Very large	6	60.0	3	30.0	1	10.0	10	100.0
Total	29	12.9	111	49.6	84	37.5	224	100.0

small enterprises were established during the colonial period. The majority were established in the 1980–93 period. The very large firms are older, many of them having been established before independence. Although there is no information on the initial sizes of these enterprises, it is likely that they exploited unique growth opportunities created by tariff protection and market distortions generated by administrative controls. The finding that smaller firms are younger and large firms older suggests a positive relationship between firm size and age. We explore this relationship in detail in Chapter 10.

The growth of micro and small-scale manufacturing enterprises in Kenya is largely a post-1960s phenomenon. One factor behind this growth has been a progressively more favourable official attitude towards the informal sector following the International Labour Organization (ILO) mission (ILO, 1972). This mission highlighted the importance of the informal sector in creating employment and incomes for Kenyans. Other factors that helped to draw more attention to the small-scale enterprises include the low employment capacity of the large-scale sector, the ongoing retrenchment programmes in both public and private sectors and broader changes in the structure of opportunities for gainful employment in Kenya. The existence of small enterprises established in the pre-independence period suggests the presence of circumstances that box enterprises into remaining small, such as peculiar response to the risk environment.

4.2.3 Enterprise ownership

The race factor

Many of the micro and small-scale firms are owned by Kenyans of African origin (Table 4.3). However, the share of African-owned businesses falls sharply as that of Asian ownership increases up the enterprise size scale. As a consequence, Asians own a majority of the medium- and large-scale enterprises. However, the ownership of the very large enterprises is more evenly

Table 4.3 Enterprise ownership by ethnic origin

	African		*Asian*		*European*		*Total*		*Not indicated*	
	No.	*%*	*No.*	*%*	*No.*	*%*	*No.*	*%*	*No.*	*%*
Micro	51	86.4	7	11.9	1	1.7	59	100.0	0	0.0
Small	26	60.5	15	34.9	2	4.6	43	100.0	2	10.0
Medium	10	15.8	51	81.0	2	3.2	63	100.0	5	25.0
Large	3	8.9	30	88.2	1	2.9	34	100.0	8	40.0
Very large	2	40.0	2	40.0	1	20.0	5	100.0	5	25.0
Total	92	45.1	105	51.5	7	3.4	204	100.0	20	

Source: Computed from the 1993 wave.

distributed. European ownership is very small for all size categories. The large-scale firms are the exception, where Europeans hold a 20 per cent share.

The concentration of African entrepreneurs in micro and small enterprises is partly explained by African entrepreneurs' limited ability to mobilise financial and human resources. African manufacturers also generally head younger firms and frequently use the informal sector as an entry point. About two-thirds of the African-owned enterprises were sole proprietorships, with only a small proportion of the enterprises being either partnerships or limited liability companies. On the other hand, more than two-thirds of the Asian-owned enterprises were limited liability companies and only a small percentage of these were either partnerships or sole proprietorships. The European-owned firms were equally distributed between limited liability and sole proprietorships (Table 4.4). To the extent that the size of an enterprise is a response to risks, African entrepreneurs seem to face peculiar risks that force them not only to start and remain small but also to operate within the simpler sole proprietorship legal status.[1]

There are also locational and sectoral differences in the legal status of enterprises (Tables 4.5 and 4.6). Eldoret, one of the small industrial towns, has the largest proportion of sole proprietorships, while Mombasa has the largest proportion of partnerships and limited liability companies. Eldoret

Table 4.4 Legal status of firms by ethnic ownership

	Not indicated		*African*		*Asian*		*European*		*Total*	
	No.	*%*	*No.*	*%*	*No.*	*%*	*No.*	*%*	*No.*	*%*
Corporation	1	5.0	0	0.0	1	1.0	1	14.2	3	1.3
Limited liability	12	60.0	17	18.5	75	71.4	3	42.9	107	47.8
Partnership	3	15.0	12	13.0	19	18.1	0	0.0	34	15.2
Sole proprietorship	1	5.0	62	67.4	10	9.5	3	42.9	76	33.9
Other	3	15.0	1	1.1	0	0.0	0	0.0	4	1.8
Total	20	100.0	92	100.0	105	100.0	7	100.0	224	100.0

Source: Computed from the 1993 wave.

Table 4.5 Legal status of small firms (0–50 employees) by location

	Nairobi		*Mombasa*		*Nakuru*		*Eldoret*		*Total*	
	No.	*%*	*No.*	*%*	*No.*	*%*	*No.*	*%*	*No.*	*%*
Sole proprietors	39	50.0	15	45.5	8	50.0	9	64.3	71	50.4
Partnerships	17	21.8	10	30.3	2	12.2	2	14.3	31	22.0
Limited	22	28.2	8	24.2	6	37.5	3	21.4	39	27.6
Total	78	100.0	33	100.0	16	100.0	14	100.0	141	100.0

Source: Computed from the 1993 wave.

is the fastest-growing town in Kenya and has very many young firms. Since debutants are likely to be local, sole proprietorships that are easier to establish are likely to dominate. The metal sector has the highest proportion of partnerships, while the food sector has the lowest proportion. It is unclear why the metal-working sector has the highest concentration of partnerships except that metal-working firms are relatively more capital-intensive and therefore require pooling of resources.

The ownership pattern shows strong participation by a small Asian minority with less than 1 per cent of the population but a large share of manufacturing enterprises (Table 4.7). An analysis of changes in the ethnicity of owners shows a rising share for Africans. Without size differentiation, the share of the African-owned enterprises rose from 7 per cent for those started during colonial period to 67 per cent for those started during the 1963–80 period. However, if micro and small-scale enterprises are excluded, the overall share of African-owned enterprises falls sharply. In other words, whereas Africans have ventured into manufacturing, not only have their operations been restricted to the small, informal end of the sector, but also some of the initially large African enterprises have exited from manufacturing altogether. Thus measures to increase participation of Africans in manufacturing will have to focus on strategies for helping not only small-scale

Table 4.6 Legal status of small firms (0–50 employees) by sector

	Food		*Textiles*		*Wood*		*Metal*		*Total*	
	No.	*%*	*No.*	*%*	*No.*	*%*	*No.*	*%*	*No.*	*%*
Sole proprietors	10	35.7	21	53.8	22	53.7	18	54.5	71	50.4
Partnerships	5	17.9	12	30.8	9	22.0	5	15.2	31	22.0
Limited	13	46.4	6	15.4	10	24.3	10	30.3	39	27.6
Total	28	100.0	39	100.0	41	100.0	33	100.0	141	100.0

Source: Computed from the 1993 wave.

Table 4.7 Evolution of ethnic ownership of enterprises

	Before 1963		*1963–80*		*1980–93*		*Total*	
	No.	*%*	*No.*	*%*	*No.*	*%*	*No.*	*%*
Not indicated	6	20.7	11	9.9	3	3.5	20	8.9
African	2	6.9	34	30.6	56	66.7	92	41.1
Asian	19	65.5	63	56.8	23	27.4	105	46.9
European	2	6.9	3	2.7	2	2.4	7	3.1
Total	29	100.0	111	100.0	84	100.0	224	100.0

Note: Column headings refer to year the firm started.
Source: Computed from the 1993 wave.

Table 4.8 Ownership structure by race

	Not indicated		*African*		*Asian*		*European*		*Total*	
	No.	*%*	*No.*	*%*	*No.*	*%*	*No.*	*%*	*No.*	*%*
Not indicated	2	10.0	0	0.0	0	0.0	0	0.0	2	0.9
Local	1	5.0	89	96.7	83	79.0	6	85.7	179	79.9
Foreign	12	60.0	0	0.0	1	1.0	0	0.0	13	5.8
Local/foreign	1	5.0	3	3.3	21	20.0	1	14.3	26	11.6
Government	4	20.0	0	0.0	0	0.0	0	0.0	4	1.8
Total	20	100.0	92	100.0	105	100.0	7	100.0	224	100.0

Source: Computed from the 1993 wave.

enterprises to graduate to medium and large enterprises, but also for ensuring sustainability of larger African enterprises. This is important in promoting a more inclusive society that is less prone to racial tensions, which in turn could undermine investor confidence.

The foreign ownership factor

The majority of the manufacturing enterprises are locally owned and the rest are either under joint local/foreign ownership, wholly foreign owned or government owned. Most of the Asian-owned enterprises are local. However, some are owned jointly by local and foreign investors (Table 4.8). The majority of the European firms are also local and the rest are under joint ownership. Considering the technological, productivity and efficiency impacts of foreign participation, Kenya's industrial policy needs to target foreign participation and encouragement of joint ownership. So far, efforts in this direction have not been particularly successful.

The gender factor

Kenyan manufacturers are predominantly male. However, the male domination varies across sectors. Other countrywide studies have shown that although female ownership is marginally greater when different sectors are pooled together, there are important sectoral variations in the sense that while males dominate manufacturing and services, small-scale commerce is largely a female domain (Table 4.9).

Splitting small-scale manufacturing into different activities confirms male domination, particularly in metal fabrication, wood and furniture making. Female ownership is therefore only more evident in the textile and garments sector (Table 4.10).

4.2.4 Entrepreneurs' background

The occupational background of the parents, especially fathers, was an important factor influencing the probability of becoming an entrepreneur.

Table 4.9 Ownership of micro and small enterprises by gender

Type of owner	*Manufacturing*	*Commerce*	*Services*	*Total*
Male	46.7	33.2	70.3	42.5
Female	39.7	53.1	13.9	43.4
Groups	13.6	13.7	15.8	14.1
Total	100.0	100.0	100.0	100.0

Source: Gemini (1994).

Table 4.10 Gender distribution of ownership of small-scale manufacturing by sector

Sector	*Male*		*Female*	
	No.	*%*	*No.*	*%*
Food	22	92.0	2	8.0
Textile	27	71.0	11	29.0
Wood	39	97.5	1	2.5
Metal	33	100.0	0	0
Total	121	89.6	14	10.4

Source: Gemini (1994).

The fathers of more than half of the entrepreneurs of the smaller firms had been either manufacturers or traders. Other studies have shown that the fathers of a third of the entrepreneurs of micro and small-scale enterprises owned manufacturing concerns, while a fifth were traders (Gemini, 1994). The parents' influence was particularly strong in the metal sector, where the fathers of more than 45 per cent of the entrepreneurs were manufacturers, and 16 per cent had either commercial or trading backgrounds. The fathers of slightly more than one-fifth of the entrepreneurs were farmers, and 13 per cent were public servants. Business heritage is therefore important in encouraging entrepreneurship. On this account, an opportunity structure that pushes Kenyans into business is likely to have a positive intergenerational impact on the business community.

The larger cities of Mombasa and Nairobi had relatively younger entrepreneurs compared with the smaller towns of Eldoret and Nakuru. In terms of sectors, textile and metal fabrication activities had more youthful entrepreneurs relative to wood and food sectors. Young school leavers, who are initially attracted to the large towns of Nairobi and Mombasa by the prospect of finding formal sector jobs, eventually get pushed into entrepreneurship, often within the informal micro and small-scale textile and metal sectors. In these, sectors the skill needs and management requirements are less exacting.

4.2.5 Enterprise specialisation

A 1993 nationwide survey of micro and small enterprises revealed that only a quarter of business owners were solely dependent on business incomes, while 70 per cent depended on businesses for at least half of their incomes. The survey also showed that entrepreneurs in urban areas relied more on incomes earned from businesses than rural-based entrepreneurs, and that female-owned enterprises provided a lower share of household income compared to male-owned enterprises (Gemini, 1994, p. 8). Many of the small-scale manufacturers in Kenya therefore engage in a wide range of other activities. Diversification is an attempt to spread risks but squanders the opportunity to develop business expertise, dissipates managerial capability, and predisposes enterprises to poor performance. The high mortality rate within micro and small-scale enterprises is partly an outcome of this entrepreneurial diversification. Improving the commercial environment would obviate some of the diversification and offer micro and small-scale enterprises an opportunity to specialise and thrive.

4.3 Technological capability

Improvements in enterprise performance and welfare depend on successful exploitation of more efficient technologies. Improvement in the mechanisms for transferring technology and in technological competence are important in promoting industrial activities. Since most African countries lack adequate capacity for meeting scientific and technological requirements, efficient access and use of imported technology becomes critical in developing an appropriate scientific and technological infrastructure (Teitel, 1993; Departments of Economics, 1994 and 1995).

In developing countries, such as Kenya, technological efforts are directed towards research and development, in-house and external training, accessing of technical support services provided by third parties, and direct purchases of technology in disembodied forms (such as foreign and domestic licences) or in embodied forms (such as foreign or domestic capital and material inputs). The stock of technical know-how generally accumulates slowly through continuous learning and innovation. The technological learning curve is a function of experience in production. There may also be technical externalities from one firm to another or from one product line to another. We can anticipate differences in access to technology, quality of technology and technological capability between different industrial activities (or sectors) and within similar industrial activities (sectors), and differences between foreign and national firms that affect productivity.

4.3.1 Foreign licences and technical assistance

Licence agreements relate to know-how for implementation of productive processes, granting of rights (often exclusive) for patent exploitation, technical assistance, blueprints, and use of brand names. Licensing agreements are concluded as part of joint venture agreements between domestic enterprises and foreign technology suppliers. A study by Seyoum (1990) on technology licensing in East Africa showed that many licensing agreements fail to provide for improvements of design and engineering capability, limiting the ability of local licensees to adapt the technology to local conditions and to innovate in other ways. The limited use of foreign licences may also be attributed to the relatively low level of foreign ownership, and contractual provisions inhibiting technological self-reliance.

Foreign licensing is not a common means of obtaining industrial technology in Kenya. Only 4 per cent of the firms reported the use of foreign licences in 1993. This percentage was even lower in subsequent years. The metal sector reported the greatest use of foreign technology licences in 1993 and was the only sector from which a few firms negotiated the use of such licences in 1994. It is perhaps in the metal-working activities where certain levels of technological sophistication are desirable. Otherwise, Kenyan manufacturers generally use simple technologies and manufacturing activities are kept in place by an accommodating local market.

The use of technical assistance is more common than foreign technology licences. Even then, there are sectoral variations as the metal and food sectors use such assistance more intensely than other sectors. This use was also greater in the more recent period, perhaps in response to the improved business prospects following liberalisation of the economy (see Chapter 2). Since technical assistance personnel are part of the foreign package, the majority of the firms reporting technical assistance also report use of foreign licences. Foreign expertise is deployed in administering firms, establishing training programmes, providing design and engineering services and in financial and marketing services. Only a few firms incur expenditure on technology licences and technical assistance. Where such expenditure is incurred, the amounts are also small.[2]

Thus not only is the incidence of the use of foreign resources and technical assistance low, but the expenditure on such licences is also minimal. Licences and technical assistance have very limited technological impact on Kenyan manufacturing. Many manufactures are of modest quality and are produced for the domestic market using simple technology.

Very few firms employ expatriate general, production or technical managers (Table 4.11). Metal-working firms engage a higher proportion of expatriate managers than firms in other sectors. In other words, not only do metal-working firms hold the greatest number of foreign licences and receive the most extensive technical assistance, they also employ the largest proportion of expatriate general and production managers. This reiterates an earlier observation that metal-working activities are more technologically exacting than other activities. Although the Kenyan education system produces relatively few technical graduates, very few firms engage

Table 4.11 Employment of expatriates, 1993–95

	1993				*1994*				*1995*			
	Foreign	*Kenyan*	*All*	*% of all*	*Foreign*	*Kenyan*	*All*	*% of all*	*Foreign*	*Kenyan*	*All*	*% of all*
General Manager												
Food	3	22	47	6	1	36	51	2	2	31	47	4
Wood	7	16	57	12	3	35	59	5	4	23	57	7
Textiles	8	16	59	14	1	30	49	2	4	17	59	7
Metal	10	20	55	18	5	35	57	9	5	22	55	9
Total	28	74	218	13	10	136	216	5	15	93	218	7
Production Manager												
Food	3	15	47	6	2	13	51	4	4	9	47	9
Wood	5	11	57	9	1	11	59	2	3	9	57	5
Textiles	7	10	59	12	4	5	49	8	3	6	59	5
Metal	7	8	55	13	5	6	57	9	3	5	55	5
Total	22	44	218	10	12	35	216	6	13	29	218	6

expatriates. We surmise that firms must be satisfied with the local personnel, or do not perceive managerial constraints as major impediments to their operations, or cannot afford foreign expertise.

The decision to hire expatriates is influenced by broad considerations such as expatriate salaries, government labour policies, expatriate qualifications, ability to train local staff, and technological constraints. Contractual arrangements with expatriates are for short periods (1–5 years) but renewable through mutual agreement. Such renewals are only necessary until there is effective transfer of technical skills and know-how to the local enterprises or until suitable local personnel become available. It has been noted that the transfer of technical know-how to Kenyan counterparts from expatriate personnel remains unsatisfactory (Tostenson and Scott, 1987).

4.3.2 Education and skills of managers

The capability stock of an enterprise lies in managerial and technical knowledge and skills. The quality of knowledge and skills available to a firm is critical for performance. High level managerial, administrative and technical personnel are required to modernise industry and the rate of such modernisation is associated with the stock and rate of accumulation of capability.

There were more general managers than production/plant managers during each of the three survey waves. The larger and more complex firms have organisational systems with hierarchical levels of authority and responsibility and relationships between different departments. For the micro and small formal and informal firms, however, the owner-managers perform most of the management functions. Given the limited specialisation among African enterprises, the owner-manager management mode is likely to be an impediment to enterprise performance and growth.

A large proportion of general managers has secondary education (Table 4.12). A similarly large proportion of production managers had either secondary or technical university education. There was more postgraduate training among production managers than general managers. The managers of large firms are also more educated than those for smaller firms. The average number of years of service for general and production managers was about 10 years. However, there are significant sectoral differences in the length of service. General managers in the wood sector had longest experience while those in the textile sector had the shortest experience. By contrast, production managers in textile had the longest experience. Furthermore, the general managers in the formal sector had nearly twice the length of service compared to those in informal firms. These differences in length of service of management in the formal and informal sector mirror differences in the life expectancy of firms in the two sectors.

Table 4.12 Education of managers, 1995

	Primary	*Secondary*	*University-NT*	*University-T*	*Postgraduate*	*% of all*
General Manager						
Food	13%	24%	11%	9%	2%	59
Wood	12%	19%	3%	10%	0%	45
Textiles	6%	19%	4%	6%	4%	38
Metal	19%	11%	4%	15%	2%	50
Total	12%	18%	6%	10%	2%	48
Production Manager						
Food	0%	9%	2%	11%	2%	24
Wood	3%	10%	0%	7%	0%	21
Textiles	2%	8%	0%	10%	0%	19
Metal	0%	2%	0%	11%	2%	15
Total	1%	7%	0%	10%	0%	20

Note: Not all firms have general or production managers, and this explains why the percentages do not add up to 100 per cent.

4.4 Market structure and competition

Kenyan manufacturing has been inwardly oriented, so that the main source of competition has been domestic. Many firms are weakened by their small size, low productive technology and commonplace activities that box them into a more competitive environment than the large ones. On the other hand, some large firms enjoyed monopolistic power in their respective markets due to government protection. The indicators of market power include excess capacity, barriers to entry due to legal restrictions, patents, production technology and large initial capital requirements and expansion problems for small firms.

One-half of the firms had more than 10 competitors for their principal product (Table 4.13). However, there are notable size and sector differences in the intensity of competition. Wood-working firms face more intense competition than firms in other sectors. Competition was also more intense for micro and small-scale firms. Large firms face only limited competition. Clearly there is an important size factor in the degree of competition as the smaller enterprises operate in a more competitive environment than the larger ones. The large enterprises therefore show costs and other advantages that secure their market shares. Although most of the competition is from domestic private firms, increasing imports are also a threat to domestic manufacturers.

The competition faced by the small firms is from other small firms. This finding is common to all enterprise sizes. Firms thus tend to compete among equals. But competition is dynamic, with regular entry and exit of

Table 4.13 Competitors for the principal product across sectors and sizes, 1994–95 (%)

	Sectors					
	Food	*Wood*	*Textiles*	*Metal*	*Total*	
None	4	3	2	3	3	
1–5	34	22	34	38	33	
6–10	20	18	12	12	16	
>10	42	58	52	47	48	
	Firm size					
	Micro	*Small*	*Medium*	*Large*	*Very large*	*Total*
None	1	2	2	8	12	3
1–5	26	30	37	38	69	33
6–10	12	17	16	24	19	16
>10	61	51	45	30	0	48

Table 4.14 Number of new and exiting important competitors by firm size and sector, 1994 and 1995

	Sector				
	Food	*Wood*	*Textiles*	*Metal*	
Entry	0.256	0.269	0.685	0.306	
Exit	0.256	0.270	0.688	1.521	
Net entry	0	−0.001	0.003	−1.215	
	Firm size				
	Micro	*Small*	*Medium*	*Large*	*Very large*
Entry	0.68	0.755	0.352	0.208	0.199
Exit	0.56	0.758	0.354	0.209	0.200
Net entry	0.16	−0.003	−0.002	−0.201	−0.001

important competitors. More food and textile firms report entry of new important competitors than firms in other sectors. However, few firms in these sectors reported the exit of important competing firms, so that there was a net increase in the intensity of competition. Similarly, most small firms report entry of new important competitors. They also report the exit of important competitors.

Although the data suggests a net decrease in the number of important competitors in the metal sector and among the large firms, this does not necessarily imply a reduction in competition since inefficient producers may have been replaced by more efficient ones (Table 4.14). Generally, there was more movement in the smaller-scale end of manufacturing than among the larger firms.

None of the firms reported entry of important state-owned competitors. This is expected since at the time of the RPED survey the government was under pressure to privatise stated-owned enterprises (Table 4.15). Nearly all the new competitors were either domestic producers or local and foreign importers. New competition from domestic producers is a bigger challenge for smaller firms, while large firms face new competition from domestic importers and foreign subsidiaries. Small firms mostly produce for the lower end of the market dominated by cheap, lower quality goods for low-income consumers, whereas the larger firms tend to produce mostly standardised and up-market products and are therefore more exposed to competition from imports and foreign subsidiaries. Textile firms are most severely affected by competition from importers due to increased importation of new or used clothes following trade deregulation. But for many of the micro and small-scale firms, competition is evidently fierce whatever their activity.

Enterprise concentration is somewhat time-variant. The results over the three years do not show any consistent differences among sectors. In 1993, concentration was highest in the textile sector where three-quarters of the sector's output was contributed by the five largest enterprises (Table 4.16).[3] In 1994, the food sector showed the greatest concentration, while the wood sector took the lead in 1995. Generally there is very low internal competition in Kenyan manufacturing.

Market stratification is evident as the micro and small-scale enterprises are exposed to intense competition, while the large enterprises bask in their extensive market power. Kenyan manufacturing exhibits significant market power and limited internal competitive pressure except perhaps at the micro and small-scale end of the sector. This is probably the result of

Table 4.15 Reported source of new important competitors by sector and firm size 1994 (%)

	Sector					
		Food	*Wood*	*Textiles*	*Metal*	*Total*
State		0	0	0	0	0
Local private firms		49	31	33	0	35
Foreign subsidiary		4	2	0	0	1
Importing firms		6	6	8	4	6
	Firm size					
	Micro	*Small*	*Medium*	*Large*	*Very large*	*Total*
State	0	0	0	0	0	0
Local private	40	41	32	30	0	35
Foreign subsidiary	0	2	0	4	0	1
Importing firms	4	4	11	7	0	6

Table 4.16 Output shares of the largest five enterprises and Herfindal indices (%)

Sector	*Firm 1*	*Firm 2*	*Firm 3*	*Firm 4*	*Firm 5*	*Concentration ratio*	*Herfindal indices (five firms)*
Food							
1993	14.20	13.00	11.60	10.40	10.40	59.60	0.091
1994	50.29	8.80	6.25	6.01	5.01	76.36	0.279
1995	21.20	14.23	9.26	7.46	6.40	58.55	0.098
Textiles							
1993	42.40	12.30	7.50	6.90	5.20	74.30	0.213
1994	37.36	21.40	9.64	7.47	3.86	79.73	0.163
1995	34.64	15.94	7.22	5.31	5.07	68.18	0.164
Wood							
1993	29.10	17.70	10.00	5.20	3.90	65.90	0.137
1994	28.20	21.30	17.07	7.42	5.16	79.15	0.205
1995	30.65	20.46	10.19	9.50	4.77	75.57	0.170
Metal							
1993	36.20	15.00	10.40	6.30	5.00	72.90	0.177
1994	21.58	15.70	12.73	9.06	8.09	67.16	0.168
1995	32.81	8.69	8.58	7.33	7.17	64.58	0.146

Sources: Economics Departments (1994, 1995); Kimuyu (1998).

years of state control and hidden entry and exit barriers. The removal of such controls will not only increase exposure to external competition, but also promote productive efficiency.

4.4.4 Disputes in input and product markets

Payment disputes are common, with some firms encountering payment problems many times in a year. Non-payment was less prevalent than late payments and the incidence of payment problems was highest in the wood and textile sectors (Table 4.17). The likelihood of enterprises suffering from payment problems was also higher for large enterprises. Formal enterprises were also more likely to complain about payment problems than the informal ones (Kimuyu, 1997). Payment-related problems occur because customers are either unable or unwilling to pay. In a few cases, payment disputes are occasioned by bankruptcy and the death of client.

Enterprises also faced disputes arising from either late or non-delivery or poor input quality. These problems were, however, less common than those related to payment. Poor input quality was more common than either late delivery or non-delivery of inputs (Table 4.18). Metal-working, wood and food processing enterprises were more prone to input problems. The supply of inputs to the wood and food sectors is more dependent on natural conditions and can therefore be expected to experience input supply difficulties during certain seasons.[4] Like payments, larger enterprises are more inclined to suffer from input problems than the smaller ones. Formal

Table 4.17 Incidence of payment problems

	Late payments		*Non-payment*	
	N	*Nr*	*N*	*Nr*
Food	29	9.68	18	3.87
Textiles	34	18.48	21	2.99
Wood	36	3.38	21	1.64
Metal	33	5.10	20	1.40
All	132	12.13	80	2.75

N = Number of firms that answered the question
Nr = Average number of late payments and non-payments respectively
Source: Economics Departments (1993).

Table 4.18 Incidence of input-related problems

	Non-delivery/Late delivery		*Poor input quality*	
	N	*Nr*	*N*	*Nr*
Food	20	5.65	27	6.5
Textiles	20	6.77	28	0.91
Wood	23	4.08	33	4.54
Metal	22	13.59	23	2.06
All	85	8.34	111	3.02

N=number who answered the question
Nr=average number of non-delivery/late delivery or poor input quality
Source: Economics Departments (1994).

enterprises were, however, less likely to suffer from input problems than informal ones probably on account of the uncertain legal status of informal enterprises. Most of the inputs used in the metal-working sector are imported either directly or indirectly. Reliability of import sources depends on availability of foreign exchange, which was a major problem during the first RPED wave.

Faced with payment problems, many of the enterprises simply wait patiently and do nothing about the dispute. Some, however, will negotiate with the offending parties, go to court, seek legal advice or use a private arbitrator (in that order) The large enterprises are more keen to use the legal machinery than others, and seldom choose to sit back and wait. By contrast, the micro and small-scale enterprises seldom press their claims, perhaps because for such enterprises the costs of taking legal action far outweigh the expected benefits. The inclination for enterprises to press for their claims following input problems was even less common. Faced with these disputes, many firms simply wait. In only a few cases are private arbitrators used or reporting to the police threatened. Use of courts and hiring of lawyers is also uncommon. Furthermore, in only a few of the payment and input cases are the disputes ever satisfactorily settled. A high propensity for corporate disputes and an inability to successfully pursue redress suggests wide opportunities for contesting business rights. This distorts the smooth pursuit of corporate interests and the development of Kenya's private sector.

4.5 Summary

The Kenyan manufacturing sector is not homogeneous. The numerous micro and small-scale enterprises in Kenyan manufacturing are fairly recent or have a low life expectancy. Many of the African-owned manufacturing firms operate as sole proprietorships, often defined around family membership. They are also smaller than those owned by other ethnic groups. Eldoret, the fastest growing town in Kenya, has attracted a large share of manufacturing new firms. The smaller firms are relatively young.

Given that sole proprietorships are particularly dominant among African manufacturers, there is very little pooling of resources and management capability. As a strategy against risks, many of the small manufacturers are diversified. This strategy dissipates managerial capability and denies such manufacturers the benefits of specialisation. But business heritage is important in Kenyan manufacturing, so that entrepreneurship that revolves around manufacturing is likely to become intergenerational.

Foreign participation is limited, denying Kenyan manufacturers the technological, productivity and efficiency externalities associated with such participation. Although about a third of the textile firms are female-owned, Kenyan manufacturing is male-dominated. In the creative and efficient use of resources, including trained personnel, management is probably the most important factor. Owner-management, which is evident among the small manufacturing sector, is likely to be an impediment to Kenya's industrial growth.

Technological competence has to be increased to facilitate technological acquisition and adaptation. This will involve building a network of applied research, development and design centres, with extension and supporting services, the effectiveness of which depends largely on good communications within the network and externally with productive sectors and government planning. While there may be limitations to adaptation and development, there is evidence of scope, in certain industries, to broaden the technological spectrum; but in that process, an educated society is critical. For this reason, industrial policies interface with those targeted at the development of human capital.

Competition is higher among smaller firms than larger firms. The small firms only compete against themselves and vice versa. Most competition is from local firms. Competition affects different sectors differently, and there are significant exit–entry size differences, with more movement at the small-scale end of the sector. By contrast, large enterprises enjoy considerable market power, which needs to be addressed in order to increase market efficiency. The opening of the economy to external competition is an effective way to deal with this problem.

Whereas enterprises are prone to payment and input related disputes, such proneness is size- and sector-specific. Opportunities for their successful resolution are limited and business transactions contestable by implication. Improving contract enforcement and the rule of law will be critical for further enterprise development.

Notes

1. The relationship between ethnicity and productivity is discussed in Bigsten, Kimuyu and Lundvall (2000).
2. Only eight firms reported that they used technology licences in 1993. The numbers were even smaller for subsequent years.
3. The concentration ratios are computed using a simple formula: let q_{ij} be the value of output produced by enterprise i in sub-sector j and Q_i the value of the sub-sector's total output, then the relative share of that enterprise can be expressed as $S_{ij} = q_{ij}/Q_j$. If all enterprises are the same size, then it is the case that $S_{ij} = 1/N_j$ where N_j is the sub-sector's sample size. The Hirschman–Herfindal index is the sum of the square of these shares and therefore has a value restricted to (0, 1) and is therefore larger when enterprise sizes are unequal.
4. The seasonality of quality and reliability of supply is likely to reflect the underlying seasonality of important segments of the road network in Kenya.

References

Bigsten, A., Kimuyu, P. and Lundvall, K. (2000), *Informality, Ethnicity and Productivity. Evidence from Small Manufacturers in Kenya*, Working Paper in Economics No. 27, Department of Economics, Göteborg University.

Coughlin, P. and Ikiara, G.K. (1991), *Kenya's Industralization Dilemma*, Heinemann, Nairobi. Department of Economics, University of Göteborg and University of Nairobi (1993), *Regional Programme of Enterprise Development: First Report on Round 1 of the Kenya Survey*, World Bank, Washington, DC.

Economics Departments, Universities of Göteborg and Nairobi (1994), *Limitations and Rewards in Kenya's Manufacturing Sector: A Study of Enterprise Development*, World Bank, Washington, DC.

Economics Departments, Universities of Göteborg and Nairobi (1995), *Manufacturing in Kenya under Adjustment*, World Bank, Washington, DC.

Gemini (1994), 'Micro and Small Enterprises in Kenya: Results of the 1993 National Baseline Survey', Gemini Technical Paper No. 75.

ILO (1972), *Kenya: Employment, Incomes and Equality*, ILO, Geneva.

Kenya (1997), *Economic Survey*, Nairobi.

Kimuyu, P.K. (1997), *Enterprise Attributes and Corporate Disputes in Kenya*, Discussion Paper No. 1, Institute of Policy Analysis and Research, Nairobi.

Kimuyu, P.K. (1998), *Industrial Policies for the 21st Century: Productivity, Competitiveness and Export Participation*, Presented at the Institute of Policy Analysis and Research National Seminar on Strategic Development Paths for Kenya in the 21st Century Safari Park Hotel, Nairobi, 24–25 April 1998.

Lall, S. (1987), *Learning to Industrialize: The Acquisition of Technological Capability by India*, Macmillan, London.

Seyoum, B. (1990), *Technology Licensing in Eastern Africa: A Critical Exposition and Analysis*, Avebury, Aldershot, England.

Teitel, S. (1993), *Technology Acquisition, Operation and Development in Selected Kenyan Manufacturing Establishments*, RPED, The World Bank.

Tostensson, A. and Scott, J.G. (1987), Kenya: Country Study and Norweigan Aid Review, The Christian Michelsen Institute, Bergen.

5
Wage Determination in Kenyan Manufacturing

Damiano Kulundu Manda

5.1 Introduction

In the last three decades, several studies (e.g., House and Rempel, 1976a, 1976b; Bigsten, 1984; Neitzert, 1992) have been undertaken in Kenya on various aspects of earnings determination, but there are still many issues that are not yet addressed. One such issue that we focus on in this chapter is the effect of firm size on wages. In particular, we analyse the effect of firm size on wages, and examine the extent to which competing theories can explain the observed effect. We address several key questions concerning the relationship. What is the effect of firm size on wages? Is the firm size–wage effect significant? Does this effect vary for different classification of workers (e.g., white-collar and blue-collar workers, men and women) and by sector? How much of the firm size–wage effect can be explained by differentials in quality of workers and compensation for unfavourable working conditions across firms? How much of the effect can be attributed to unions and firm behaviour?

Previous empirical studies (House and Rempel, 1976a; Brown and Medoff, 1989; Evans and Leighton, 1989; Gerlach and Schmidt, 1990; Teal 1995; Velenchik, 1997a) report a positive relationship between firm size and wage rates. This implies that workers in large firms earn higher wages than their counterparts in small firms. The firm size–wage differentials between large and small firms are in some cases large and significant.[1] Most of the empirical studies on the effect of firm size on wages have been undertaken in developed countries; there have been very few in developing countries, including Africa (e.g., House and Rempel, 1976a; Teal 1995; Velenchik, 1997a). Lack of data on both firm and individual characteristics in the developing world partly explains this state of affairs. The RPED survey on the Kenyan manufacturing sector, which is used in our analysis, fills that data gap by generating information on both individual and firm characteristics.

The survey was conducted over 1993–95, a period of rigorous implementation of structural adjustment programmes and, therefore, the analysis is likely to provide some evidence of the short-term responses of the labour market to a set of policy changes during the period.[2] For instance, some large firms (especially in the textile sub-sector) declined due to competition following trade liberalisation. Also, real wages fell steeply in the period 1990–94 and unemployment was high (see Chapter 2). Equally important, an analysis of the effect of firm size on wages can give an indication of the functioning of the labour market. If labour markets were competitive, then individual and firm characteristics that are unrelated to productivity or to workers' utility would have no influence on wages. In Kenya, real wages have fluctuated over time and considerable microeconomic evidence suggests that skills are correctly priced. For example, Knight and Sabot (1990) showed that the much faster increase in supply of educated labour force in Kenya than in Tanzania accounted for the fall in returns to education in Kenya. Therefore, the existence of firm size–wage differentials could be evidence that the labour market is not perfectly competitive.[3]

The role of micro and small enterprises in job creation has increased in Kenya following the implementation of structural adjustment programmes, which started in the early 1980s. For example, 20 per cent of the total manufacturing sector employees are employed in firms with up to five employees, 24.8 per cent are in firms with 6–75 employees, 18.9 per cent are in firms with 76–500 employees and 36.2 per cent are in firms with more than 500 employees (Departments of Economics, 1994). Firms employing up to five employees form a majority of the firms in the manufacturing sector. An analysis of the effect of firm size on wages is important for understanding the wage levels for workers in the micro and small enterprises and wages for workers in larger enterprises, and the process of wage determination in all enterprises. Thus the analysis of firm size–wage effect can have important implications for income distribution.

The remaining part of the chapter is organised as follows. Section 5.2 reviews the theoretical explanations of the effect of firm size on wages. Section 5.3 presents the model, and describes data and variables. The results are presented in section 5.4. Finally, the conclusion is in section 5.5.

5.2 Theoretical explanations for firm size–wage differentials

Both neoclassical and institutional labour economists provide explanations as to why large employers pay more than small employers. Alternative explanations are based on firm behaviour and efficiency wage theories. The neoclassical labour economists base their explanation on the theory of 'equalising differences' or 'compensating differentials' and emphasise the firm size differentials in labour quality and working conditions between large and small firms. The institutionalist explanations are more varied, but emphasise the effect of trade unions and product market power.

According to the labour quality explanation, the firm size–wage differentials can be explained by differences in the quality of workers employed in small and large firms. Large and small firms pay the same wage for workers of a given quality, but differences in wages are due to the fact that large firms employ higher quality workers. Large firms' operations are often more capital-intensive than small firms and may involve the use of specialised equipment; they therefore have the incentive to employ higher-quality workers, probably due to capital–skill complementarity. Also, Oi (1990) points out that large firms may use higher quality labour to reduce monitoring costs.

The labour quality explanation can be tested statistically by estimating a wage equation controlling for measured labour quality using education and experience as proxies. The problem is that the omission of unmeasured labour qualities such as ability can lead to biased estimates. Different approaches have been used in previous empirical studies to overcome the problem caused by unmeasured labour quality. One approach uses proxies for ability, such as IQ in a cross-section wage equation that includes firm size (Blackburn and Neumark, 1992). Another approach involves the use of longitudinal data and takes the first-difference estimates on a full sample or separately on individuals who have worked for different sized employers and those who have not changed employers (Evans and Leighton, 1989).[4] This approach reduces the bias on the effect of firm size on wages caused by the omission of unmeasured labour quality (assuming ability is time-invariant).

The working condition explanation is based on the undesirable conditions at the place of work. According to the theory of compensating wage differentials, workers in jobs with undesirable characteristics (e.g., dangerous working environment) should be compensated with higher wages.[5] Studies on firm size–wage effect in developed countries give several reasons why working conditions in large firms could be undesirable. These are that large work places are associated with undesirable conditions such as greater reliance on rules, less freedom of action and scheduling (Masters, 1969) and a more impersonal working atmosphere (Lester, 1967). If large firms compensate workers for such undesirable working conditions, then wages for workers in large firms may be higher than for workers in small firms. However, this may not be valid for a country such as Kenya with many unemployed job seekers. In any case the working conditions in small firms may be inferior to those in large firms. For example, labour standards and health and safety regulations are more likely to be adhered to in large firms than in small enterprises.

To test for this explanation, one needs to isolate the unattractive working conditions at the work place. Controlling for variables capturing these working conditions in a wage equation can reduce or eliminate the wage premium associated with firm size if most of the premiums are due to compensating differentials.

Institutional economists emphasise the role played by trade unions in wage determination. If trade unions can mobilise many employees effectively, the joint action by these employees can affect the wage level. Trade unions can also raise the wage level through collective bargaining. If the size–wage effect is due to 'strong' unions which can raise wages, we would expect to find more variations in wages for union workers than for non-union workers. This is partly due to there being little pressure for employers to increase wages for non-union workers and partly because union threats are likely to be greater in large firms that often have higher proportions of union workers than small firms. For example, Freeman and Medoff (1981) found a positive relationship between union wages and the proportion of organised labour, but no relationship between non-union wages and the proportion non-union workers. In Kenya union membership is voluntary and about 30 per cent of formal sector employees are union members. Estimates based on RPED data show that about 33 per cent of employees in the manufacturing sector are union members. More than half of the workers in firms with over 50 employees are union members, while less than 20 per cent of the workers in firms with up to 50 employees are union members. The importance of the effect of trade unions on the firm size–wage effect can be investigated by, for example, determining whether firm size–wage effect exists for non-union and union workers.

The neoclassical and institutional theories presented above so far assume cost-minimisation behaviour. However, some firms (especially large firms) may find it optimal to pay workers wages that are higher than the market-determined wages. This behaviour can be rationalised by assuming that some firms find that increasing wages above the market wage is beneficial (Krueger and Summers, 1988). Efficiency wage theories provide explanations as to why some firms pay wages that are higher than the market determined wages.

The first explanation is that some firms pay higher wages to reduce turnover (Stiglitz, 1974). Reducing turnover is more important for large firms because their operations may involve special equipment that require specific handling skills and, therefore, have the incentive both to invest in training and to protect the investment with a wage premium if the skills acquired are not firm-specific (Collier and Gunning, 1999). Second, higher wages may raise the workers' effort by raising the cost of job loss, thereby reducing shirking. Finally, firms may raise wages to attract higher quality workers. This is desirable especially when quality is not directly observable.

Another explanation is that wage determination is characterised by rent-sharing so that wages depend upon the employers' ability to pay, and more profitable firms pay higher wages. One common argument is that large firms with monopoly power may share with their workers some excess profits or rents that such power yields. The source of the rents to the manufacturing sector in Kenya may be the government policy that restricted both local and foreign competition.[6] To the extent that large firms are more profitable, possibly because they enjoy large rents from protection, workers in larger firms can exact a premium (Velenchik 1997b; Collier and Gunning, 1999).

Even if firms do earn rents it is not clear why they would choose to share them with employees. A possible explanation is that rent-sharing is the outcome from bargaining process, in which the relative bargaining power of firms and workers determines the distribution of the rents. However, given the widespread unemployment in Kenya, particularly for less skilled labour and weak trade unions, it is unlikely that most workers have any meaningful bargaining power. Velenchik (1997a, 1997b) argues that real bargaining may be between firms and the government since firms recognise that their access to the rents is mainly dependent on government policy and so they share some rents with workers in an attempt to keep the government happy. In this chapter we attempt to test the extent to which these theories can explain the effect of firm size on wages.

5.3 The model, data and variables

The empirical analysis of the effect of firm size on wages is based on the ordinary least square (OLS) estimation technique.[7] We estimate the following wage equation:

$$w_i = X_i\beta + e_i$$

where w_i is log of hourly wages for individual i (i=1,2,...,n), X is a vector of explanatory variables, β is a vector of parameters to be estimated and e_i is the error term. In analysing the effect of firm size on wages, time dummies are included to capture changes in wages over time. To examine the importance of firm size in explaining wages, we estimate several variants of the wage equation controlling for the various theoretical explanations using proxy variables.

The labour market sections of the RPED survey provide information on both firm and individual characteristics. The information on firms was elicited from representatives of each firm and includes total number of employees, total wage bills, profits and firm ownership, among other things. This provides more accurate information on, for instance, firm size than would be the case when individuals provide such information. The information on individual characteristics of the workers was obtained through interviews with at most ten employees randomly chosen from a list of workers at each firm giving adequate representation of different categories of workers. The data provides information on, for example, education level, previous experience, tenure, age, sex, hours of work, wages, and union status. On average, about two employees were interviewed in firms with up to five employees, five in firms with 6–50 employees and eight in firms with more than 50 employees. The total sample used in our analysis is 3506 observations obtained by pooling the data on individuals over time

(1206 individuals in 1993, 1146 in 1994 and 1154 in 1995). The sample excludes individuals for whom some information was missing and individuals below 18 years and above 64 years. The data is pooled to increase the number of observations (e.g., the female sample is small in each year).

The variables used in the analysis are defined in Table A5.1 in the Appendix. The dependent variable is the natural logarithm of the individual hourly wages deflated by the consumer price index (CPI), with 1986 as the base year. The hourly wages were derived by dividing monthly wages by the total number of hours of work in a month (i.e., hours of work in a week multiplied by 4.3 weeks in a month).

To analyse the effect of firm size on wages one of the explanatory variables in the wage equation should be a measure of firm size. Firm size is measured in terms of the number of employees in a firm. Firm size can be measured either as a continuous variable in the estimation or by using dummy variables to denote the observed size classes. We use both measures of firm size in our analysis of firm size–wage effect. For the continuous measure of firm size, we use the natural logarithm of the number of workers in a firm that has been used in several studies such as Brown and Medoff (1989) and Velenchik (1997a).[8] Five firm-size dummy variables are created by classifying the firms into five size groups as follows: micro firms (1–5 workers), small firms (6–50 workers), medium firms (51–100 workers), large firms (101–500 workers) and very large firms (more than 500 workers).[9] Some firms declined and others expanded during the period, but most of them fall within the same classification. If a firm grows during the period, for example, its number of employees increases from four in 1993 to 10 in 1994, then it is classified as a micro firm in 1993 and as a small firm in 1994. However, this applies to very few firms (about 12 per cent) in our sample.

The dummy variables for location (3), sub-sectors (9), education level (4), on-the-job training (1), outside training (1), occupation (7), sex (1), union status (1), permanent employment (1), weekly hours of work greater than 50 (1), shifts (1) and time (2) are included to capture the effects of the variables on wages.[10] On-the-job training can be taken to be firm specific, while outside training (done outside the firm) may include vocational training or other forms of training done outside the firm that may be assumed to be general to all firms usually. The time dummy variables are included to capture the time effects on wage rates that arise due to changes in macroeconomic factors and in policies.

The other variables of interest include the following: first, tenure with the current employer and its square. Previous research on effect of firm size on wages provides at least two reasons to expect a steeper tenure profile in large organisations. The first emphasises the effect of tenure profiles on retention and employee diligence, while another places the focus on the internal markets accessible to employees of large organisations. The motivation for employers adopting steeper tenure profiles in the first case includes specific training (Oi, 1990) and costs for monitoring employee performance (Lazear, 1981). The effects of tenure and its square on wages are expected to be positive and negative respectively. If tenure explains part of the size–wage effect, then the high wage paid by large firms is partly an effort to reduce turnover.[11]

Second, previous experience, (i.e., the number of years an individual worked elsewhere before he or she started working for the current employer) is included; the usual measures of experience, number of years in employment and age, are excluded because they are highly correlated with tenure. Lastly, profit per worker is derived by dividing annual profits for each firm by the total number of workers in the firm. Profit per worker for each firm was included not only because it shows the rents available for the firm to share with the employees, but because it may also be a measure of the firm's ability to pay. Its effect on wages is expected to be positive. When estimating the effect of profit on wages, we exclude individuals from firms that do not provide data on profit. This affects individuals from about 30 firms that had missing data on profits.

5.4 Results

5.4.1 Firm size and wages

We begin presenting our results by examining the relationship between firm–size and wages. The firm size effect on wages is analysed for the whole sample and separately for various sample groups defined in terms of broad occupational categories (white-collar and blue-collar), sub-sectors, and sex (males and females).[12] This is done to enable us to compare the outcomes of the various groups. We begin by presenting the descriptive statistics of the distribution of the hourly wages by firm size.

The means and standard deviations of the individual hourly wages for the various sample groups by firm size are presented in Table 5.1 (the means of all explanatory variables are presented in Table A5.2 in the Appendix). As shown in the table, the hourly wage rates increase with firm size. The hourly wages for workers in firms with up to five employees are about half of the wages for individuals in firms with more than 100 employees.[13] The average hourly wages for males and females increase with firm size, but females earn less than males in firms with up to 50 employees and higher than males in firms with more than 50 employees. For each firm size, the hourly wages for white-collar workers are higher than for blue-collar workers. The wages for white-collar and blue-collar workers are higher in larger firms than in smaller ones.

There are some variations in the distribution of wages by sub-sector, but generally the hourly wages increase with firm size. However, in the textile sector the hourly wages for workers in very large firms are lower than for

Table 5.1 Real hourly wages for various groups by firm size (1986 = 100)

Groups	*Micro*	*Small*	*Medium*	*Large*	*Very large*
All workers	4.8	6.0	7.9	9.7	10.5
	(3.0)	(4.9)	(7.4)	(8.2)	(10.3)
Males	5.0	6.2	7.9	9.6	10.3
	(3.1)	(5.2)	(7.6)	(8.2)	(10.6)
Females	3.9	5.2	8.1	10.2	11.5
	(2.4)	(2.8)	(6.6)	(8.3)	(8.4)
White-collar	5.4	7.8	12.1	13.1	16.0
	(2.5)	(8.0)	(11.4)	(9.3)	(15.5)
Blue-collar	4.7	5.8	7.2	8.5	8.9
	(3.0)	(4.2)	(6.2)	(6.7)	(7.6)
Textiles	5.0	5.6	6.2	8.7	7.6
	(3.5)	(4.5)	(5.9)	(7.4)	(5.5)
Food	3.6	6.0	6.4	10.2	10.3
	(2.0)	(6.3)	(6.5)	(9.6)	(7.7)
Wood	5.5	5.5	9.0	10.0	14.4
	(3.2)	(3.7)	(7.5)	(8.5)	(11.7)
Metal	4.7	7.2	9.7	9.8	15.3
	(2.5)	(5.6)	(8.5)	(6.8)	(12.8)

Note: The standard deviations are in the parentheses. Wages are in natural logarithms.
Source: Calculation based on RPED data pooled over time.

workers in large firms. This can be explained by the fact that some of the very large textile firms sampled are public companies, which were facing both financial and management problems during the survey period. The average hourly wage for individuals employed in micro firms in the food sector is the lowest in the four sectors. The standard deviation of the hourly wages increases with firm size. This shows that pay differences are higher among workers in the larger firms than in the smaller firms.

The OLS parameter estimates of the wage equations for various sample groups are presented in Table A5.3 in the Appendix. The estimated effect of firm size on wages for the various groups is reproduced in Table 5.2 below. The sample of individuals is drawn from firms that vary largely by, for example, type of production, skill requirements, technology and so on which could lead to heteroscedasticity. A test for heteroscedasticity using White's test shows that it is a problem. This means that the parameter estimates are consistent, but the standard errors are not. We use White's consistent parameter covariance matrix to estimate consistent standard errors.[14] We also test for pooling data over time. The F-test rejects pooling across time period since the calculated F is significant at the 0.01 level (F-calculated = 11.4 and F-critical = 1.46). We, therefore, allow variations across time by including time dummies.

As shown in Table 5.2, row 1, the estimated coefficients of firm size for the whole sample are positive and significant at the 0.01 level. For the firm

Table 5.2 Estimates of firm–size effect on wages for various sample groups

	Group	*Log (size)*	*Small*	*Medium*	*Large*	*Very large*
1	All	0.1010**	0.1482**	0.3060**	0.4091**	0.5121**
		(13.247)	(3.373)	(5.735)	(7.967)	(7.415)
2	Men	0.0940**	0.1368**	0.2688**	0.3833**	0.4607**
		(11.357)	(2.959)	(5.695)	(7.584)	(7.059)
3	Women	0.1413**	0.1854	0.5029**	0.5056**	0.7626**
		(7.251)	(1.677)	(4.328)	(4.218)	(5.210)
4	White-collar	0.1041**	0.0765	0.3154*	0.3917**	0.4237**
		(5.131)	(0.568)	(2.198)	(2.742)	(2.665)
5	Blue-collar	0.0936**	0.1542**	0.2894**	0.3842**	0.4857**
		(12.080)	(3.440)	(6.076)	(7.815)	(7.625)
6	Textile	0.0820**	0.0998	0.1950*	0.4326**	0.3815**
		(4.857)	(1.399)	(2.537)	(5.435)	(3.446)
7	Food	0.1246**	0.2854**	0.3335**	0.6068**	0.6490**
		(7.466)	(2.669)	(2.954)	(5.188)	(4.474)
8	Wood	0.0898**	0.0434	0.2443**	0.2954**	0.3767**
		(6.220)	(0.579)	(2.816)	(3.515)	(3.514)
9	Metal	0.1149**	0.2302**	0.4901**	0.5145**	0.7249**
		(7.602)	(2.867)	(5.587)	(5.009)	(6.516)

* Significant at 0.05 level.
** Significant at 0.01 level.
F-test that the firm size wage differentials jointly equal to zero rejects at 0.01 level (F-cal. ranges from 9.16 to 44.9) and the F-critical = 3.32. The control variables include 3 location dummies, sex dummy, union dummy, time dummies, education dummies, the training dummies, tenure and tenure squared and previous experience.
Note: t-statistics are in the parentheses.

size dummies, the estimates show that the hourly wage rates for workers in firms with more than five employees are significantly higher than the hourly wage rate for workers in firms with up to five employees. The magnitude of the firm-size coefficients increases with the firm size indicating that the wages for workers in larger firms are higher than in smaller firms. For example, the hourly wages for workers in small, medium, large and very large firms are 16, 36, 51 and 67 per cent higher than the hourly wages for workers in micro firms.[15] The estimates are lower than those reported by Velenchik (1997a) on Zimbabwe, but higher than those reported in studies carried out in developed economies.[16]

Rows 2 to 9 of the table show the estimated coefficients of firm size effect on wages for various sample groups. The estimates show that the hourly wages for both men and women working in firms with more than five employees are significantly higher (except for women in small firms) than the hourly wages for men and women working in firms with up to five employees. However, the relative pay for women in firms with more than 50 employees seems higher than for men.

The wages for white-collar and blue-collar workers in firms with more than five employees are significantly higher than for those working in firms with up to five employees. The results also show that hourly wages in the food, textile, wood and metal sub-sectors increase with firm size, but there are differences in the magnitude and significance of the coefficients across the four sub-sectors. The firm size–wage effect seems higher in the metal and food sectors than in the other two sectors. The log (size) coefficient is positive and significant for all the sample groups.

Before concluding this section we briefly comment on the result of other variables included in the analysis. The discussion on these variables is based on model 1 (see the Appendix) estimated for the entire sample, but we also highlight the results of the various groups.

The hourly wage rates for workers in firms in Nakuru, Eldoret and Mombasa are significantly lower than for workers in firms in Nairobi. The result seems consistent across various sample groups. There are different possible explanations for the result. First, it is most likely that firms with a higher ability-to-pay are located in Nairobi because most government activities and skilled labour are concentrated in Nairobi, and Nairobi has a well-developed infrastructure compared with other urban centres and is easily accessible to the rest of the world. Second, the higher wage in Nairobi may be due to the high cost of living in Nairobi compared with other urban centres because of rapidly increasing population. The high cost of living may be used as a justification for higher wage demands.

The four education dummy variables have the expected positive sign showing that the hourly wages for workers with primary, secondary, technical and polytechnic, and university education levels are higher than for individuals without education. The education dummies are significant except the primary dummy variable and the secondary dummy variable for women and white-collar workers. The coefficients of the education dummy variables increase with education level showing that the hourly wage for individuals with higher education level is higher than for those with lower education level. It seems education is insignificant in determining wages in the wood sector, but it is significant in the food and metal sectors, while in the textile sector only technical education is significant at the 0.05 level.

The hourly wage rate for individuals with on-the-job training and outside training is significantly higher than for individuals without on-the-job training and outside training, respectively. For the whole sample and the sample groups, the relative pay for individuals with on-the-job training is lower than the relative pay for individuals with outside training.

Tenure and tenure squared have the expected positive and negative signs respectively. This shows that wage increases as tenure increases, but at a decreasing rate. The peak earnings are attained at 27.8 years of working for the same employer. The effect of tenure on hourly earnings of white-collar and female workers is, however, not significantly different from zero at the

0.05 level. The returns to previous experience are positive and significant. This implies that workers with a longer previous working experience earn higher wages with the current employer (firm) than workers with a shorter previous working experience.

The 1994 time dummy variable is negative, while that for 1995 is positive and both are significant at the 0.01 level. The coefficients of the time dummies show that the hourly wage level in 1994 was 23.6 per cent lower than the hourly wage level in 1993, while in 1995 it was 9.7 per cent higher than the 1993 level. An F-test that the time dummies are jointly equal to zero is rejected at the 0.01 level (F-cal. = 98.35 and F-crit. (2, 3484) = 4.61). The high decline in 1994 was partly due to the high inflation and low economic activities in the period 1991 to 1993. The increase in hourly wage level in 1995 reflects the recovery of the economy due to economy-wide and labour market reforms that included the relaxation of wage guidelines in 1994, allowing for adjustments in wages to take into account the full cost of living and productivity of the workers.

Workers in the formal sector firms receive significantly higher wages than workers in the informal sector. On average, the hourly wages for workers in the formal labour market firms are 20.4 per cent higher than the hourly wages for those in the informal sector. Union workers' wages are 28.8 per cent lower than that for non-union workers.

In summary, the analysis done in this section provides evidence of a strong positive relationship between firm size and wages. For instance, employees in firms with more than five workers earn wages that are 16–67 per cent higher than employees in firms with up to five workers. These differences persist even in the presence of control variables that would be expected to capture most of the cross-firm differences in labour quality. The positive effect of firm size on wages seems robust. Women earn higher hourly wages than men in firms with more than 50 employees and lower than men in firms with up to 50 employees. For each firm size, the hourly wage for white-collar workers is higher than for blue-collar workers.

Can the effect of firm size on wages be explained by the various competing theories? We address this question in the next section. The methodology used in our analysis is to introduce variables that serve as proxies for the theories in the wage equation and examine whether the firm size coefficients decline, and whether they remain significant or become insignificant. A reduction in the coefficient of the firm size effect on wages would suggest that the proxies included in the equation are part of the explanation. If the firm size effect remains significant, then the interpretation is that other factors besides the proxies included are also important in explaining the size–wage effect.

5.4.2 Differential labour quality explanations

The labour quality explanation asserts that workers in large firms are highly skilled and, therefore, the wage premium in large firms is a reward

Table 5.3 Alternative degrees of controls for labour quality and the firm size–wage effect

	Controls	*Log (size)*	*Small*	*Medium*	*Large*	*Very large*
1	3 location dummies, sex dummy, union dummy, 2 time dummies, formal sector dummy.	0.1251** (16.11)	0.1796** (4.195)	0.3712** (8.008)	0.4994** (10.628)	0.6447** (10.203)
2	Row (1) controls plus 4 education dummies, On-the-job training dummy & outside training dummy, Previous experience	0.1100** (14.53)	0.1672** (3.901)	0.3378** (7.350)	0.4434** (9.424)	0.5632** (9.332)
3	Row (2) controls plus tenure and tenure squared	0.1010** (13.35)	0.1482** (3.473)	0.3060** (5.735)	0.4091** (7.967)	0.5121** (7.415)
4	((row1 − row3) /row1)*100	19.2%	17.5%	17.6%	18.1%	20.6%

* Significant at 0.05 level.
** Significant at 0.01 level.
F-test that the firm size-wage differentials jointly equal to zero rejects at 0.01 level. (F-calculated ranges from 44.9 to 68.9, and F-critical = 3.32).
Note: t-statistics are in the parentheses.

for their high productivity. As shown in Table 5.3, the proxies used to control for measured labour quality (i.e., education, training, tenure and previous experience) cannot fully explain the firm size effect on wages.

To examine the extent to which the proxies for labour quality explain the effect of firm size on wages, we explore the impact of alternative degrees of controls of human capital on the firm size–wage effect. The results are presented in Table 5.3. If the firm size–wage effect is due to differences in measured labour quality across firms, we would expect a substantial decline in the coefficients of firm size when we control for measured labour quality.

Row 1 in Table 5.3 shows the estimates of a model based on the entire sample excluding the labour quality proxies, (i.e., education, on-the-job training, outside training, tenure and tenure square, and previous experience). Then additions of the human capital factors are made, as shown in rows 2 and 3. As can be seen from row 4 of the table, the human capital proxies of the measured labour quality included in the regression reduce the log (size) coefficient by 19.2 per cent and the coefficients of the firm size dummies by an average of 18.5 per cent. The figure is the mean of the figures in the last four columns of row 4. However, the firm size coefficients are still significant, indicating that the measured labour quality proxies included in the equations do not fully explain the effect of firm size on wages.

It is possible that the human capital variables may not adequately control for labour quality. The reason could be that the omitted unmeasured labour quality differences in ability and motivation vary systematically across firms and are being 'picked up' by the firm size rather than the human capital controls. The effect of unmeasured labour quality can be eliminated by estimating a first difference wage equation. Unfortunately, we cannot apply this method to the data that we have and it is hard to tell the extent to which unmeasured labour quality explains the firm size–wage effect. However, previous studies that have estimated a first difference wage equation (e.g., Brown and Medof, 1989; Evans and Leighton, 1989) show that unobserved labour quality explains a considerable portion of the size-wage effect but does not eliminate the effect.

We end this section with an investigation of whether there are any major differences in employee characteristics between the different firm sizes and an analysis of returns to human capital factors in each firm size. There are several reasons why one would suspect extensive sorting of heterogeneous workers across heterogeneous firms.[17] The first is that smaller firms are less capital intensive than larger firms and, since capital and skill are complements, one would expect large firms to have a larger proportion of better quality workers. The second is that small firms are more likely to fail and have more variable growth than large firms. This means that jobs at smaller firms are of shorter and more variable duration than jobs at larger firms.

Some differences in the workers' characteristics (in terms of education) exist across the firm sizes. The proportions of workers without education and those with primary level education are higher in small firms than large firms, while the proportion of workers with secondary education level increase with firm size (see Table A5.2 in the Appendix). The proportions of workers with technical and polytechnic, and university education are higher in large and very large firms than in smaller firms. The share of workers with some training is also higher in large firms than small ones. It appears that larger firms recruit better-quality workers than the small firms.

We estimated a wage equation for each firm size (results can be obtained from the author). The returns on education for workers in the micro firms are lower than for workers in the other firms. The education coefficients in the micro firms are insignificant, showing that education is not an important determinant of wages in the micro firms. The returns on technical and polytechnic, and university education increase with firm size. The returns on training (especially outside training) are also higher in large firms than in small firms. This shows that large firms pay higher wages for a given labour quality than the smaller firms, probably to attract the best workers.

5.4.3 Compensating differential explanations

The measured labour quality proxies reduced the firm size coefficient by less than 20 per cent. In this section, we examine the importance of compensating differentials in explaining the firm size–wage differentials. Thus we address the question of whether the pay premium in the large firms is a reward for undesirable working conditions.

To provide a direct and precise answer to this question is not possible because undesirable working conditions are difficult to define and hard to measure. Due to these problems the only feasible approach used in many empirical studies is to conduct several less direct, but reasonable analyses which, when taken together, can permit judgement on the compensating wage differential explanation of the firm size–wage effect. We make attempts similar to those made in previous empirical studies to explain the firm size–wage effect by controlling for undesirable working conditions in the wage equation. The first of these attempts examines the extent to which the firm size coefficient is affected by taking detailed controls for industries and occupations in a wage equation.[18] We divided the textile, food and wood sectors into much more detailed segments.[19] We then estimated the wage equation including nine industry and seven occupation dummies in addition to the labour quality controls in Table 5.3, row 3.

The estimated firm size effect on wages is shown in Table 5.4 (the rest of the estimated results are in Table A5.4 in the Appendix). As shown in row 2

Table 5.4 Firm size effect on wages with controls for working conditions

	Controls	*Log (size)*	*Small*	*Medium*	*Large*	*Very large*
1	Same as row 3 Table 5.3	0.101** (13.35)	0.1482** (3.373)	0.3060** (5.735)	0.4091** (7.967)	0.5121** (7.415)
2	Row 1 controls plus 9 industry dummies, 7 occupation dummies	0.098** (11.71)	0.1451** (3.429)	0.2929** (6.435)	0.3597** (7.725)	0.4251** (7.212)
3	Row 3 controls plus 1 weekly hours dummy, non-permanent employment and shifts dummy	0.092** (10.75)	0.1220** (3.086)	0.2788** (6.424)	0.3537** (7.765)	0.4247** (6.987)
4	((row1 − row4)/ row1)*100	9.0%	17.7%	08.9%	13.5%	17.0%

* Significant at 0.05 level.
** Significant at 0.01 level.
F-test that the firm size-wage differentials jointly equal to zero rejects at 0.01 level.
Note: t-statistics are in the parentheses.

of Table 5.4, including the industry and occupation dummies in the wage equation reduces the coefficient of firm size slightly, but they are still significant at the 0.01 level. A possible explanation could be that most of the firm size–wage differentials occur within very well detailed industries and occupations, but are not captured by differences in working conditions across the industry and occupation proxies.

We then included the job conditions identified in the literature as indicators of unattractive working conditions (e.g., long hours of work, shifts and non-permanent employment).[20] The result after adding the three variables shows that the coefficients of firm size decline as shown in row 3, Table 5.4. Overall, the proxy variables for working conditions reduce the log (size) coefficient by 9 per cent and firm size dummy coefficients by 14.0 per cent (derived by calculating the mean of the last four columns of row 4). However, the firm size effect is still significant at the 0.01 level. This shows that most of the firm size–wage effect is not due to differences in rewards for undesirable working conditions.[21]

5.4.4 Tenure and firm size

The relationship between tenure and firm size can give more insight to the positive relationship between firm size and wages. On average, tenure increases with firm size. For instance, the average tenure for micro, small, medium, large and very large firms is 4.5, 7, 9, 9 and 10 years, respectively (see Table A5.2 in the Appendix). The tenure for workers in micro firms is about half that for workers in very large firms. A possible explanation for the longer tenure in large firms could be the higher pay in larger firms than in smaller ones. If the high wage in large firms is the only reason for the positive firm size–tenure relationship, then controlling for the wage rate in a regression of tenure on firm size will eliminate the effect. The estimated results of a regression of tenure on firm size, log hourly wages, schooling, and so on are shown in Table 5.5.[22]

The results show that years of tenure with an employer grow significantly with firm size independent of the firm size–wage effect. Thus the coefficients of firm size are positive and significant at the 0.01 level and increase with firm size, indicating that tenure is higher in larger firms than in smaller ones. The possible reasons why tenure may be higher in large firms include, first, the possibility of greater opportunities with a large employer (probably due to existence of internal labour markets in large firms) for workers to develop their careers and move from one assignment to another within the firm.[23] (However, if a worker does not like to work in a large firm *per se*, he or she can still quit).

The second reason is the lower failure probability for large firms compared with small ones (see Idson, 1996; Evans and Leighton, 1989). The failure probability for small firms in Kenya is high because of problems such as loss of work place, adverse political and personal circumstances,

Table 5.5 Determinants of tenure (model 21)

Variables	*Coefficients*	*t-statistic*
Small firms	0.0932*	2.066
Medium firms	0.1930**	3.592
Large firms	0.1980**	3.576
Very large firms	0.3322**	4.549
Nakuru	−0.0059	−0.159
Eldoret	0.0533	1.227
Mombasa	0.0843*	2.383
Sex	0.1217**	3965
Primary	−0.3620**	−5.492
Secondary	−0.6933**	−10.78
Technical and polytechnic	−0.7603**	−9.669
University	−0.8715**	−10.99
On-the-job training	0.0126	0.412
Outside training	−0.0780*	−2.468
Other experience	−0.1656**	−8.000
1994	0.0354	1.086
1995	−0.1183**	−3.533
Food sector	0.0680*	2.003
wood sector	0.1096**	3.260
Metal sector	0.0374	1.121
Union status	0.2435**	9.162
Log hourly wages (predicted)	0.1617**	7.579
Constant	0.3684**	5.596
Adj-R	0.174	
F-statistics	33.00**	
N	3506	

* Significant at 0.05 level.
** Significant at 0.01 level.
Dependent variable is years of tenure.

inadequate supplies, intense competition that limits profits, low demand and small initial capital (McCormick, 1993; Ongile and McCormick, 1996; McCormick, Kinyanjui and Ongile, 1997). The higher failure probabilities for smaller firms may lead to an increase in resignations as workers search for alternative employment in anticipation of job loss.

Third, large firms can invest in higher levels of employee on-the-job training and will pay higher wages and fringe benefits to reduce turnover.[24] This is supported by the evidence that wages for workers with on the on-the-job training are higher than for workers without. The coefficients of the education dummies are negative and significant and their size increases with education level, indicating that the highly educated are less likely to stay with one employer for a longer time than the less educated. Union workers seem to have a higher tenure with the employer than non-union workers. A possible explanation for this could be that firms often try to avoid the confrontation with the trade unions that arises from dismissals. This could be true given the evidence that most of the trade disputes in Kenya are related to dismissals.[25]

5.4.5 The union threat explanation

Labour unions are one institutional force that can influence the level of wages. In particular, where unions can mobilise most employees effectively, the joint action of these employees can affect wage levels and working conditions via the threat of strikes. However, the law limits the extent to which the threat of strikes can be used in Kenya, and legal strikes are rare.[26] The unions are also financially weak and cannot provide financial assistance to the strikers. Nevertheless, union power exists in the manufacturing sector to some extent because manufacturing labour is one of the best organised in Kenya (Chege, 1988). Also, unions can still have some influence on wages through the collective bargaining process.

The analysis of the effect of unions on the firm size–wage effect requires the assessment of the union and non-union sectors separately. In this section we assess the union effect by estimating a wage equation separately for union and non-union workers. If firm size–wage differentials are due to varying degrees of union power across firm size, we would expect to find less variation in wages for non-union workers across firms. The estimated firm size coefficients after controlling for human capital factors are presented in Table 5.6.

Rows 1 and 2 show the results for non-union workers in all firms and in firms with no union workers, respectively. The results in the two rows show that the coefficients of firm size for non-union workers are positive and significant at the 0.01 level. The estimated coefficients increase with firm size, indicating that the hourly wage for non-union workers is higher in larger firms than in smaller firms. Row 3 shows that the firm size–wage effect for union workers is insignificant. Compared with the micro firms, the premium in firms with more than five employees ranges from 18.0 per cent to 88.1 per cent for non-union workers and from −6.5 per cent to 14.6 per cent for the union workers. The wage premium is even higher for non-union workers in firms without union workers, as shown in row 2. In this case the premium ranges from 19.1 per cent to 171.9 per cent. Thus, non-union workers have higher and significant firm size-wage differentials than union workers. Union strength is not likely to be part of the explanation of the firm size–wage effect in Kenya.

It is likely that the high firm size–wage effect for non-union workers is due to the firms' ability to pay. A study by House and Rempel (1976b) shows that unionisation is not stronger in industries that have above-average productivity and profit per worker. We investigate the effect of firms' ability to pay on the firm size–wage effect in the next section.

Table 5.6 Estimates of the size–wage effect for union and non-union members

Variables	*Non-union workers*		*Non-union workers in firms with no union workers*		*Union workers*	
	Coefficient	*t-statistic*	*Coefficient*	*t-statistic*	*Coefficient*	*t-statistic*
Small firms	0.1653**	3.623	0.1752**	3.988	−0.0672	−0.697
Medium firms	0.3828**	7.306	0.3966**	6.823	0.0190	0.194
Large firms	0.4623**	8.614	0.5212**	7.424	0.1360	1.371
Very large firms	0.6319**	8.477	1.0001**	3.681	0.1118	1.045
Nakuru	−0.4859**	−12.167	−0.4782**	−10.062	−0.3300**	−6.799
Eldoret	−0.4866**	−10.605	−0.4852**	−10.117	−0.2352**	−4.211
Mombasa	−0.2134**	−6.980	−0.2326**	−5.776	−0.0635	−1.569
Sex	0.0478	0.907	0.0594	1.286	0.1302**	3.139
Primary	−0.0406	−0.113	−0.0188	−0.258	0.0728	1.262
Secondary	0.1172	1.947	0.0610	0.832	0.1312*	2.249
Technical and polytechnic	0.2834**	5.750	0.2532*	2.517	0.1062	1.271
University	0.2714**	6.343	0.2406**	2.577	0.2983**	3.450
On-the-job training	0.0966**	2.825	0.0875*	2.318	0.0855*	2.513
Outside training	0.2157**	6.636	0.2022**	4.749	0.1994**	4.974
Tenure	0.3433**	4.835	0.3664**	5.389	0.1499**	2.732
Tenure squared	−0.0797*	−2.219	−0.0926**	−3.214	−0.0270	−1.436
Other experience	0.0672**	4.491	0.0778**	2.866	0.0165	0.710
1994	−0.2607**	−5.101	−0.2508*	−5.513	−0.3323**	−8.777
1995	0.0116**	4.805	0.0006	0.013	−0.0404	−1.018
Formal sector	0.1270**	4.025	0.1273**	2.981	0.1900**	2.566
Constant	1.2565**	14.186	1.2791**	12.943	1.2575**	10.445
Adj-R	0.277		0.282		0.187	
F-statistics	28.42**		25.82**		15.381**	
N	1428		1266		1252	

*Significant at 0.05 level.
**Significant at 0.01 level.

5.4.6 The rent-sharing explanation

The firm size–wage differentials may also be due to the behaviour of firms, (e.g., large firms with monopoly power may share their profits or rents with workers). We introduced a variable denoting the size of rents available to be shared with workers – profit per worker – in addition to the controls in row 3, Table 5.4, and estimated the equation for individuals in firms that provided information on profit only.[27] The firm–size effect on wages when profits per worker are included in the equation is presented in row 2 of Table 5.7. As shown in the table, profits have a negligible effect on the firm size–wage effect and the coefficients of firm size are still significant at the 0.01 level. However, profit has a positive effect on wages and is significantly

Table 5.7 Effect of profit per worker on the firm–size effect on wage

	Controls	*Log (size)*	*Small*	*Medium*	*Large*	*Very large*
1	Same as row 3, Table 5.4	0.089** (10.65)	0.1104** (2.545)	0.2601** (5.399)	0.3158** (6.323)	0.3870** (6.987)
2	Row 1 plus profits per worker	0.0878** (10.43)	0.1018** (2.359)	0.2358** (4.900)	0.3116** (6.275)	0.3847** (6.149)

** Significant at 0.01 level.
F-test that the firm size-wage differentials are jointly equal to zero rejects at 0.01 level (F-calculated for row 2 is 27.79, F-critical = 3.42).
Note: t-statistics are in the parentheses.

different from zero at the 0.01 level. This result is similar to that found by Velenchik (1997a), Teal (1995), and Blanchflower, Oswald and Sanfey (1996) who also find a positive relationship between profits and wages.

A possible interpretation of the positive relationship between profits and wages is that firms engage in rent-sharing in response to collective bargaining with trade unions. Given that the bargaining position of workers is weak due to weak trade unions and bargaining skills, the real bargain may be between the firms and the government, which facilitates the generations of rents by pursuing policies that protect firms from competition (see Velenchik 1997a, 1997b). It may also show that firms which make high profits have the ability to pay high wages.[28]

Alternatively, the positive relationship could be explained using the efficiency wage hypothesis. Firms may raise wages to raise potential profits by increasing productivity through increased workers' effort. Firms may also share rent with workers to reduce voluntary turnover, thereby reducing training costs. This is supported by the fact that in our sample, tenure is higher in larger firms than in smaller ones. Such a long-run relationship between large-firm employers and employees (which is common in the internal labour markets of large firms) is conducive to the payment of wages above the labour market wage to maintain the loyalty, cooperation and morale of the employees (Idson, 1996; Knight, 1998).

The positive relationship between wages and profits may also show risk-sharing (Collier and Gunning, 1999). It is argued that firms may shift some risks they face to workers, paying lower wages when profits fall. If risk-sharing is the explanation for the positive wage relationship between wages and profits, wages will be more sensitive to changes in profits than to the level of profit.[29] It is, therefore, not possible to distinguish between rent-sharing and risk-sharing on cross-section estimates presented in this chapter. However, Teal (1995) and Velenchik (1997a, 1997b) estimated both cross-section and panel estimates of the wage–profit relationship for Ghana and Zimbabwe respectively, and found that wages were more sensitive to changes in profits than to cross-section differences in the level of profits.

5.5 Conclusion

The analysis in this chapter shows that wages increase with firm size. This result holds for various groups of workers classified by occupation, gender and sub-sector. The strong positive firm size–wage effect persists even after controlling for labour quality, undesirable working conditions and profits (a measure of the amount of rents). The high pay in large firms is not due to strong trade unions.

Large firms employ better quality workers than the small firms. The returns to workers with longer previous experience, technical and polytechnic, and university education are higher in large firms than in small ones. Thus, large firms may pay higher wages partly to attract better quality labour. Tenure increases with firm size, indicating that turnover may be lower in large firms than in small ones. This shows that large firms pay higher wages to reduce turnover and, hence, reduce training costs. The long tenure in large firms could also be due to the existence of internal labour markets in the firms (allowing workers to move from one assignment to another), and low failure probabilities in large firms. Firms also share rents with the workers probably to increase potential profits through increased productivity or to maintain the loyalty, cooperation and morale of their employees.

The crucial point is that the firm size–wage differentials are large and significant after controlling for labour quality, undesirable working conditions and other institutional factors. The existence of firm size–wage differentials shows that the Kenyan labour market deviates from the competitive labour market model.

The strong positive relationship between firm size and wages implies that productive growth in firm size can lead to expanded employment and rising wages. The expanded firm employment and rising real wages rest critically on the eradication of obstacles to firm growth which include, among other factors, lack of access to credit facilities by the micro, small and medium firms.

Appendix

Table A5.1 Definition of variables

Variables	*Definition*
Firm size:	
Mirco	1 if individual works in a firm with up to five employees, 0 otherwise
Small	1 if individual works in a firm with 6 to 25 employees, 0 otherwise
Medium	1 if individual works in a firm with 26 to 75 employees, 0 otherwise
Large	1 if individual works in a firm with 76 to 500 employees, 0 otherwise
Very large	1 if individual works in a firm more than 500 employees, 0 otherwise
Location:	
Nairobi	1 if individual works in Nairobi, 0 otherwise
Nakuru	1 if individual works in Nakuru, 0 otherwise
Eldoret	1 if individual works in Eldoret, 0 otherwise
Mombasa	1 if individual works in Mombasa, 0 otherwise
Sex	1 if male, 0 otherwise
Education:	
Primary	1 if one has primary education, 0 otherwise
Secondary	1 if one has secondary education, 0 otherwise
Technical and polytechnic	1 if one has technical and polytechnic education, 0 otherwise
University	1 if one has University education, 0 otherwise
On-the-job training	1 if an individuals has received some training within firm, 0 otherwise
Outside training	1 if an individual has received training outside the firm, 0 otherwise
Tenure	Number of years an individual has worked for the current firm
Previous experience	Number of years of working before getting a job with the current firm
1993	1 if individual observed in 1993, 0 otherwise
1994	1 if individual observed in 1994, 0 otherwise
1995	1 if individual observed in 1995, 0 otherwise
Formal sector	1 if works in the formal sector, 0 otherwise
Union status	1 if individuals is a member of a labour union, 0 otherwise
A non-permanent worker	1 if one is a temporary employee, 0 otherwise
Shifts	1 if an individual works in a firm with at least two working shifts
Long hours of work	1 if an individual works more than 48 hours per week, 0 otherwise

Table A5.1 continued

Variables	*Definition*
Occupation:	
occupation1	1 if a manager, engineer, physical scientist, administrators and clerical workers, 0 otherwise
occupation2	1 if commercial and sales, 0 otherwise
occupation3	1 if supervisor and foremen, 0 otherwise
occupation4	1 if technician, 0 otherwise
occupation5	1 if an equipment maintenance worker, 0 otherwise
occupation6	1 if skilled production worker, 0 otherwise
occupation7	1 if other production worker, 0 otherwise
occupation8	1 if support staff, 0 otherwise
Sector:	
Metal	1 if individual works in the metal sub-sector, 0 otherwise
Wood	1 if individual works in the wood sub-sector, 0 otherwise
Furniture	1 if individual works in the Furniture sub-sector, 0 otherwise
Textile	1 if individual works in the textile sub-sector, 0 otherwise
Garment	1 if individual works in the garment sub-sector, 0 otherwise
Grain mills	1 if individual works in the grain mills product sub-sector, 0 otherwise
Bakery	1 if individual works in the bakery sub-sector, 0 otherwise
Sugar, confectioneries, beverages and, tobacco	1 if individual works in the sugar, confectionery, beverage and tobacco sub-sector, 0 otherwise
Meat products, oils and fats	1 if individual works in the meat products, oil and fats sub-sector, 0 otherwise
Miscellaneous food	1 if individual works in the miscellaneous food sub-sector, 0 otherwise
Profit per worker	Total annual firm profits divided by total firm workers
Hourly wages	Individual monthly wages divided by hours of work in a month

Table A5.2 The means for the variables used in the analysis (for all individuals, firm size and various group classifications)

Variables	*All*	*Micro*	*Small*	*Medium*	*Large*	*Very large*
Micro firms	0.070	–	–	–	–	–
Small firms	0.416	–	–	–	–	–
Medium firms	0.231	–	–	–	–	–
Large firms	0.230	–	–	–	–	–
Very large firms	0.053	–	–	–	–	–
Nairobi	0.657	0.505	0.557	0.557	0.865	0.546

Table A5.2 continued

Variables	*All*	*Micro*	*small*	*Medium*	*Large*	*Very large*
Nakuru	0.100	0.106	0.137	0.098	0.027	0.135
Eldoret	0.089	0.078	0.102	0.100	0.027	0.222
Mombasa	0.154	0.311	0.204	0.101	0.081	0.097
Sex	0.868	0.848	0.866	0.877	0.871	0.859
No education	0.053	0.041	0.076	0.049	0.026	0.022
Primary	0.337	0.504	0.382	0.330	0.225	0.237
Secondary	0.497	0.340	0.454	0.509	0.593	0.573
Technical and polytechnic	0.055	0.041	0.036	0.058	0.092	0.059
University	0.058	0.074	0.052	0.054	0.054	0.108
On-the-job training	0.244	0.201	0.241	0.232	0.269	0.281
Outside training	0.175	0.172	0.158	0.158	0.212	0.222
Tenure	8.075	4.487	7.510	8.900	8.900	10.210
Previous experience	5.391	5.311	5.269	5.823	5.532	5.243
1993	0.350	0.340	0.301	0.401	0.357	0.395
1994	0.321	0.320	0.341	0.306	0.351	0.205
1995	0.329	0.340	0.358	0.293	0.292	0.400
Formal sector	0.904	0.430	0.876	0.978	1.000	1.000
Union status	0.357	0.111	0.320	0.422	0.405	0.476
Non-permanent workers	0.167	0.352	0.237	0.116	0.059	0.070
Shifts	0.224	0.000	0.079	0.301	0.370	0.773
Long hours of work	0.139	0.328	0.157	0.111	0.100	0.119
occupation1	0.113	0.037	0.079	0.119	0.158	0.254
occupation2	0.043	0.061	0.058	0.030	0.032	0.011
occupation3	0.098	0.016	0.052	0.107	0.160	0.259
occupation4	0.021	0.004	0.016	0.020	0.030	0.059
occupation5	0.232	0.156	0.265	0.201	0.229	0.222
occupation6	0.275	0.463	0.275	0.299	0.229	0.119
occupation7	0.093	0.094	0.095	0.112	0.075	0.076
occupation8	0.125	0.169	0.160	0.112	0.087	0.119
Metal	0.268	0.262	0.247	0.320	0.260	0.249
Wood	0.113	0.057	0.120	0.132	0.088	0.162
Furniture	0.145	0.209	0.141	0.093	0.222	0.000
Textiles	0.086	0.086	0.022	0.112	0.150	0.179
Garment	0.127	0.238	0.179	0.108	0.027	0.091
Grain mills	0.049	0.037	0.066	0.010	0.000	0.000
Bakery	0.057	0.000	0.053	0.115	0.035	0.000
Sugar, beverage, tobacco and confectioneries	0.035	0.004	0.058	0.006	0.042	0.000
Meat products, oils and fats	0.046	0.049	0.017	0.063	0.053	0.157
Miscellaneous food	0.044	0.045	0.067	0.043	0.037	0.054
Profit per worker	32855	18053	35671	32677	37074	36967
Wages	7.457	4.799	6.033	7.934	9.662	10.49

Table A5.2 continued

Variables	*Male*	*Female*	*Textiles*	*Food*	*Wood*	*Metal*	*Non-union*	*Union*
Micro firms	0.068	0.080	0.106	0.051	0.060	0.068	0.096	0.021
Small firms	0.415	0.423	0.394	0.433	0.449	0.384	0.440	0.374
Medium firms	0.233	0.216	0.241	0.235	0.176	0.275	0.208	0.273
Large firms	0.231	0.225	0.192	0.235	0.263	0.223	0.213	0.261
Very large firms	0.052	0.056	0.067	0.046	0.052	0.049	0.043	0.070
Nairobi	0.644	0.741	0.698	0.522	0.637	0.766	0.610	0.740
Nakuru	0.104	0.078	0.071	0.127	0.138	0.061	0.113.	0.077
Eldoret	0.092	0.071	0.032	0.114	0.146	0.052	0.104	0.061
Mombasa	0.160	0.110	0.198	0.236	0.078	0.121	0.172	0.121
Sex	–	–	0.828	0.815	0.918	0.896	0.855	0.891
No education	0.056	0.020	0.061	0.075	0.053	0.026	0.050	0.058
Primary	0.353	0.233	0.379	0.309	0.386	0.279	0.316	0.375
Secondary	0.473	0.657	0.476	0.515	0.453	0.544	0.502	0.488
Technical and polytechnic	0.057	0.045	0.044	0.055	0.041	0.079	0.063	0.042
University	0.060	0.045	0.040	0.046	0.067	0.072	0.069	0.037
On-the-job training	0.248	0.220	0.214	0.248	0.249	0.260	0.235	0.261
Outside training	0.168	0.220	0.139	0.160	0.185	0.206	0.198	0.133
Tenure	8.323	6.444	7.598	8.835	8.629	7.636	6.992	10.02
Previous experience	5.697	3.378	5.945	4.566	5.637	5.444	5.236	5.670
1993	0.352	0.292	0.373	0.315	0.348	0.344	0.330	0.371
1994	0.323	0.348	0.329	0.313	0.321	0.343	0.339	0.305
1995	0.324	0.361	0.297	0.372	0.331	0.314	0.333	0.324
Formal sector	0.903	0.907	0.917	0.929	0.880	0.894	0.870	0.963
Union status	0.367	0.294	0.398	0.367	0.336	0.337	–	–
Non-permanent workers	0.173	0.130	0.160	0.156	0.195	0.155	0.220	0.073
Shifts	0.226	0.212	0.236	0.406	0.041	0.239	0.211	0.248
Long hours of work	0.147	0.095	0.124	0.225	0.086	0.130	0.175	0.073
occupation1	0.105	0.238	0.087	0.145	0.106	0.149	0.150	0.074
occupation2	0.039	0.073	0.058	0.046	0.038	0.036	0.059	0.018
occupation3	0.103	0.048	0.096	0.087	0.088	0.112	0.105	0.080
occupation4	0.024	0.013	0.016	0.016	0.015	0.041	0.024	0.022
occupation5	0.247	0.110	0.202	0.191	0.207	0.307	0.199	0.283
occupation6	0.270	0.274	0.345	0.147	0.361	0.232	0.265	0.282
occupation7	0.096	0.071	0.102	0.120	0.074	0.081	0.075	0.125
occupation8	0.116	0.173	0.094	0.248	0.111	0.042	0.123	0.143
Metal	0.257	0.192	–	–	–	–	0.263	0.222
Wood	0.121	0.065	–	–	–	–	0.101	0.135
Furniture	0.153	0.093	–	–	–	–	0.173	0.095
Textiles	0.078	0.080	–	–	–	–	0.064	0.104

Table A5.2 continued

Variables	*Male*	*Female*	*Textiles*	*Food*	*Wood*	*Metal*	*Non-union*	*Union*
Garment	0.121	0.173	–	–	–	–	0.122	0.138
Grain mills	0.044	0.080	–	–	–	–	0.053	0.042
Bakery	0.055	0.069	–	–	–	–	0.055	0.061
Sugar, beverage, tobacco and confectioneries	0.032	0.058	–	–	–	–	0.036	0.034
Meat products, oils and fats	0.041	0.076	–	–	–	–	0.045	0.047
Miscellaneous food	0.042	0.080	–	–	–	–	0.033	0.064
Profit per worker	32923	32426	22722	55472	19493	34290	35399	26781
Wages	7.498	7.190	6.317	7.096	7.551	8.593	8.097	6.307

Table A5.3 Estimate of firm size effect on wages for various groups

Variables	*Model 1# (All)*	*Model 2# (Men)*	*Model 3 (Women)*	*Model 4 (White collar)*	*Model 5# (Blue collar)*	*Model 6 (Textiles)*
Small firms	0.01482**	0.1368**	0.1854	0.0765	0.1542**	0.0998
	(3.373)	(2.959)	(1.677)	(0.568)	(3.440)	(1.399)
Medium firms	0.3060**	0.2688**	0.5029**	0.3154*	0.2894**	0.1950**
	(5.735)	(5.395)	(4.328)	(2.198)	(6.076)	(2.537)
Large firms	0.4091**	0.3833**	0.5056**	0.3917**	0.3843**	0.4326**
	(7.967)	(7.584)	(4.218)	(2.742)	(7.815)	(5.435)
Very large firms	0.5121**	0.4607**	0.7626**	0.4237**	0.4857**	0.3815**
	(7.415)	(7.059)	(5.210)	(2.665)	(7.625)	(3.446)
Nakuru	−0.4375**	−0.4330**	−0.4674**	−0.3147**	−0.4386**	−0.3906**
	(−15.49)	(−14.77)	(−5.266)	(−3.002)	(−15.440)	(−5.261)
Eldoret	−0.3857**	−0.3711**	−0.5315**	−0.3216**	−0.3818**	−0.3273**
	(−11.63)	(−9.998)	(−5.703)	(−3.335)	(−10.909)	(−2.545)
Mombasa	−0.1826**	−0.1885**	−0.1762*	−0.0715	−0.1952**	−0.2128**
	(−7.039)	(−6.712)	(−2.247)	(−0.883)	(−7.471)	(−4.480)
Sex	0.0533*	–	–	0.0797	0.0829**	−0.0248
	(2.075)	–	–	(1.271)	(2.864)	(−0.470)
Primary	0.0215	0.0171	0.0673	0.0272	0.0290	0.0725
	(0.635)	(0.406)	(0.428)	(0.143)	(0.714)	(0.895)
Secondary	0.162**	0.1261**	0.1463	0.2069	0.1019*	0.0802
	(3.370)	(2.920)	(0.935)	(1.143)	(2.456)	(0.967)
Technical and Polytechnic	0.3503**	0.3342**	0.5245**	0.5570**	0.2225**	0.2454*
	(5.951)	(5.403)	(2.760)	(2.891)	(3.533)	(2.094)
University	0.4415**	0.4327**	0.5846**	0.7613**	0.2036**	0.1648
	(6.901)	(6.352)	(3.039)	(3.978)	(2.902)	(1.366)
On-the-job training	0.0944**	0.1023**	0.0165	0.1899*	0.0799**	0.0150
	(3.869)	(3.902)	(0.257)	(2.530)	(3.162)	(0.290)
	0.2237**	0.2097**	0.2886**	0.1964**	0.2274**	0.1386**

Table A5.3 *continued*

Variables	*Model 1# (All)*	*Model 2# (Men)*	*Model 3 (Women)*	*Model 4 (White collar)*	*Model 5# (Blue collar)*	*Model 6 (Textiles)*
Outside training	(8.514)	(7.283)	(4.623)	(2.756)	(8.339)	(4.421)
	0.1984**	0.2123**	0.1729	0.1606	0.1999**	0.1727
Tenure	(5.744)	(5.724)	(1.475)	(1.146)	(5.517)	(2.209)*
	−0.0358**	−0.0381**	−0.0452	0.0012	−0.0376**	−0.0427
Tenure squared	(−3.057)	(−3.097)	(−0.879)	(0.029)	(−3.097)	(−1.444)
	0.0746**	0.0750**	0.1030	0.1230**	0.0562**	0.0110
Previous	(4297)	(4.101)	(1.914)	(2.807)	(3.209)	(0.366)
experience	−0.2340**	−0.2326**	−0.2343**	−0.0761	−0.2203**	−0.1527**
	(−8.698)	(7.983)	(−3.366)	(−1.046)	(−7.524)	(−2.980)
1994	0.0936**	0.0977**	0.0906	0.2215**	0.0791**	0.1071
	(3.238)	(3.107)	(1.237)	(2.818)	(2.523)	(1.925)
1995	0.1858**	0.1736**	0.2662**	0.2642*	0.1526**	0.1112
	(4.886)	(4.165)	(2.689)	(2.321)	(3.828)	(1.493)
Formal sector	−0.2529**	−0.2404**	−0.3167**	−0.2034**	−0.2265**	−0.2398
	(−12.86)	(−11.929)	(−5.639)	(−2.996)	(−11.449)	(−6072)
Union status	1.2160**	1.2897**	1.0609**	1.0249**	1.2281**	1.3816**
	(18.136)	(18.223)	(5.903)	(4.146)	(17.5217)	(11.980)
Constant	0.253	0.270	0.405	0.307	0.260	0.193
	60.24**	57.25**	16.71**	13.315**	49.490**	9.47**
N	3506	3043	463	585	2921	746

* Significant at 0.05 level.
** Significant at 0.01 level.
Note: t-statistics are calculated using consistent standard errors.

Table A5.3 Estimate of firm size effect on wages for various groups

Variables	*Model 7 Food*	*Model 8 Wood*	*Model 9 Metal*	*Model 10 1993*	*Model 11 1994*	*Model 12 1995*
Small firms	0.2854**	0.0434	0.2302**	0.0978	0.1580*	0.2230**
	(2.669)	(0.579)	(2.867)	(1.284)	(2.546)	(3.387)
Medium firms	0.3335**	0.2443*	0.4901**	0.2240**	0.3067**	0.4304**
	(2.954)	(2.816)	(5.587)	(2.749)	(4.514)	(5.894)
Large firms	0.6068**	0.2954**	0.5145**	0.3434**	0.4125**	0.5111**
	(5.188)	(3.515)	(4.570)	(4.016)	(6.038)	(6.915)
Very large firms	0.6490**	0.3767**	0.7249**	0.3519**	0.5299	0.6383**
	(4.474)	(3.514)	(6.516)	(3.347)	(5.326)	(7.268)
Nakuru	−0.3494**	−0.4342**	−0.5296**	−0.3377**	−0.5117**	−0.4130**
	(−5.343)	(−7.752)	(−7.297)	(−5.580)	(−10.59)	(−8.414)
Eldoret	−0.298**	−0.3721**	−0.4753**	−0.2978**	−0.4243**	−0.3908**
	(−4.290)	(−6.688)	(−6.029)	(−4.524)	(−8.282)	(−7.737)

Table A5.3 continued

Variables	*Model 7 Food*	*Model 8 Wood*	*Model 9 Metal*	*Model 10 1993*	*Model 11 1994*	*Model 12 1995*
Mombasa	−0.1324*	−0.1784*	−0.1138*	−0.1779**	−0.2120**	−0.1337**
	(−2.495)	(−2.501)	(−2.048)	(−3.414)	(−5.080)	(−3.387)
Sex	0.0797	0.0047	−0.0389	0.0472	0.0614	0.0747
	(1.505)	(0.077)	(−0.692)	(0.857)	(1.472)	(1.857)
Primary	0.0355	−0.1232	0.0499	0.0304	0.1666*	0.0199
	(0.429)	(−1.514)	(0.458)	(0.398)	(2.397)	(0.248)
Secondary	0.1946*	−0.0483	0.2036	0.1165	0.1309*	0.1817*
	(2.321)	(−0.613)	(1.901)	(1.445)	(2.173)	(2.201)
Technical and	0.3166**	0.1782	0.4842**	0.3087**	0.2265**	0.6290**
polytechnic	(2.714)	(1.607)	(3.978)	(2.977)	(2.845)	(5.770)
University	0.6459**	0.1582	0.5615**	0.6156**	0.2446**	1.1548**
	(5.281)	(1.567)	(4.583)	(5.085)	(3.364)	(9.666)
On-the-job	0.1544**	0.0712	0.0728	0.2975**	0.0810*	0.0593*
training	(2.945)	(1.607)	(1.591)	(3.203)	(2.143)	(2.059)
Outside training	0.1900**	0.2067**	0.2455**	0.1250	0.1644**	0.2473**
	(3.212)	(4.369)	(5.372)	(1.488)	(4.973)	(6.146)
Tenure	0.1692*	0.0903	0.3081**	0.0170	0.3697	0.2382**
	(2.167)	(1.429)	(4.033)	(0.244)	(6.726)	(4.064)
Tenure squared	−0.017	−0.0162	−0.0535	0.0127	−0.0872**	−0.0231
	(−0.655)	(−0.723)	(−1.773)	(0.509)	(4.521)	(1.067)
Previous	0.0728	0.0685*	0.1230**	0.0197	0.0796**	0.2200**
experience	(1.853)	(2.369)	(4.135)	(0.791)	(2.749)	(6.983)
1994	−0.2985**	−0.2239**	−0.2206**	–	–	–
	(−5.103)	(−4.584)	(−4.624)			
1995	0.0516	0.1232*	0.0705	–	–	–
	(0.865)	(2.512)	(1.352)			
Formal sector	0.3795**	0.2562**	0.0396	0.0781	0.2169**	0.1222*
	(4.088)	(4.019)	(0.564)	(1.071)	(4.133)	(2.152)
Union status	−0.2096**	−0.2369**	−0.2928**	−0.1297**	−0.3553**	−0.2926**
	(−4.728)	(−6.428)	(−7.553)	(−3.610)	(−11.09)	(−8.983)
Constant	0.7757**	1.5475**	1.2088**	1.4591**	0.9441**	1.1297**
	(5.582)	(11.677)	(8.262)	(11.79)	(10.06)	(10.74)
Adj-R	0.291	0.306	0.336	0.158	0.315	0.453
F-statistics	17.58**	21.39**	23.66**	12.90**	28.62**	51.26**
N	850	970	940	1207	1145	1154

* Significant at 0.05 level.
**Significant at 0.01 level.
Note: t-statistics in parentheses.

Table A5.4 Estimates of firm size effect on wages controlling for working conditions and profits

Variables	*Model 13#* *Coefficient*	*T-statistics*	*Model 14#* *Coefficient*	*T-statistics*	*Model 15#* *Coefficient*	*T-statistics*
Small firms	0.1451**	3.108	0.1220**	3.086	0.1018**	2.359
Medium firms	0.2929**	6.435	0.2788**	6.424	0.2358**	4.900
Large firms	0.3597**	8.083	0.3537**	7.765	0.3116**	6.275
Very large firms	0.4251**	7.556	0.4247**	6.987	0.3847**	6.145
Nakuru	−0.4139**	−13.17	−0.3881**	−12.63	−0.4008**	−12.17
Eldoret	−0.3595**	−11.13	−0.3347**	−10.52	−0.3334**	−9.925
Mombasa	−0.1462**	−5.469	−0.1592**	−6.027	−0.1533**	−5.334
Sex	0.0812**	3.013	0.1038**	3.931	0.0965**	3.437
Primary	0.0119	0.281	0.0199	0.487	−0.0028	−0.062
Secondary	0.0493	1.163	0.0585	1.412	0.0317	0.704
Technical and polytechnic	0.2119**	3.810	0.2465**	4.534	0.2249**	3.852
University	0.2963**	5.308	0.3138**	5.753	0.3009**	5.090
On-the-job training	0.0807**	3.431	0.0813**	3.540	0.0800**	3.185
Outside training	0.1903**	7.519	0.1962**	7.931	0.1914**	7.194
Tenure	0.1556**	4.424	0.1554**	4.424	0.1217**	3.229
Tenure squared	−0.0289*	−2.292	−0.0288*	−2.309	−0.0198	−1.479
Other experience	0.0530**	3.402	0.0522**	3.427	0.0634**	3.851
1994	−0.1292**	−4.701	−0.0993**	−3.691	−0.0289**	−2.562
1995	0.1643**	5.726	0.1815**	6.471	0.2079**	6.860
Formal sector	0.2015**	5.759	0.0917*	2.604	0.1506*	3.786
Union status	−0.2011**	−10.26	−0.2134**	−11.05	−0.2187**	−10.46
Occupation2	−0.2238**	−4.360	0.1896**	−3.779	−0.1842**	−3.730
Occupation3	−0.1298**	−3.176	0.1481**	−3.709	−0.1260**	−3.421
Occupation4	−0.2445**	−3.794	0.2399**	−3.813	−0.2451**	3.822
Occupation5	−0.3838**	−10.25	0.3993**	−10.87	−0.3917**	−8.470
Occupation6	−0.2873**	−8.231	0.3013**	−8.818	−0.2863**	−8.513
Occupation7	−0.2809**	−6.589	0.2970**	−7.085	−0.2771**	−6.642
Occupation8	−0.4635**	−11.87	0.4615**	−11.95	−0.4605**	−11.29
Medal sector	0.1320**	4.597	0.1367**	4.872	0.1813**	5.573
Garment	−0.0658	−1.918	−0.0574	−1.709	−0.0314	−0.832
Bakery product	−0.1774**	−3.982	−0.0574	−1.193	0.0518	0.890
Grain mill products	0.0532	1.139	0.0445	0.972	0.1541*	2.257
Dairy, canned food and oil	0.0875	1.829	0.1398**	2.981	0.2373**	4.045
Sugar, confectionery, beverage and tobacco	0.1606**	3.018	0.1391**	2.667	0.2869**	4.145
Miscellaneous foods	0.0627	1.284	0.0687	1.434	0.1789*	2.859

Table A5.4 continued

Variables	*Model 13#* *Coefficient*	*T-statistics*	*Model 14#* *Coefficient*	*T-statistics*	*Model 15#* *Coefficient*	*T-statistics*
Wood	0.0328	0.926	0.0084	0.240	0.0648	1.697
Furniture	0.0537	1.616	0.0302	0.911	0.0896*	2.429
Long hours of work	–	–	−0.0196**	−13.10	−0.0185**	−11.48
Non-permanent employees	–	–	0.0652*	2.528	0.4672	1.666
Shifts	–	–	−0.0320	−1.221	0.0001	0.004
Profit per worker	–	–	–	–	0.0011**	3.930
Constant	1.4595**	19.56	1.4566**	19.28	1.8055**	15.65
Adj-R	0.334		0.366		0.380	
F-statistics	48.58**		51.60*′		44.62**	
N	3506		5606		3060	

* Significant at 0.05 level.
** Significant at 0.01 level.
Note: #t-statistics calculated using consistent standard errors.

Notes

1. For example, Mellow (1982) reports that the average wage for workers in large firms was 12 per cent higher than in small firms in the USA (after adjusting for differences in human capital and institutional settings). Evans and Leighton (1989) and Velenchik (1997a) report much higher firm size–wage differences.
2. Rapid implementation of structural adjustment programmes took place in the period 1992–95 (Ikiara and Ndung'u, 1997). The economic reforms implemented included the removal of virtually all price and foreign exchange controls, liberalisation of domestic marketing trade, import liberalisation, reduction of the budget deficit, privatisation, relaxation of wage guidelines, and other market reforms. Most of the reforms are aimed at eliminating policy distortions and lead to a competitive labour market that generates efficient levels of employment and wages to foster economic growth.
3. The different sources for evidence that labour markets are not competitive include, for example: (a) interindustry wage differentials (Krueger and Summers, 1988); (b) employer size–wage differentials (Brown and Medoff, 1989; Idson and Feaster, 1990; Velenchik, 1997a); and (c) relationship between wages and firm profitability (Blanchflower, Oswald and Sanfey, 1996).
4. See Evans and Leighton (1989) for a discussion on effect of using this method in analysing firm size–wage effect.
5. Examples of variables used to measure undesirable working conditions include (a) irregular working hours (Kostiuk, 1990); (b) inconvenient working hours, job insecurity and poor working conditions (Mcnabb, 1989), hours constraints, hard physical work, dangerous work, and stressful work (Duncan and Holmlund, 1983). In Kenya inconvenient working hours are likely to be common in small firms where longer hours of work are common and firms where there is more than one working shift.

6. Relatively large firms may also enjoy economies of scale which, together with freedom from competition, would give them the opportunity to pay above-average wages.
7. It would have been better to use panel data estimation techniques (see, for example, Hsiao, 1986; Baltagi, 1995) to capture individual specific effect and time specific effect, but data on individuals is not a panel.
8. Brown and Medoff (1989) and Velenchik (1997a) treat firm size as continuous variable in their studies by taking the log of number of workers in a firm. However, there is no theoretical or empirical basis for imposing this particular functional form and this may obscure the true wage–size relationship.
9. This classification is similar to that used in the RPED reports for Kenya except for some changes in the classification of firms with 6–500 workers. This classification is important because it allows us to compare pay in micro and small enterprises (which form the majority of firms in the manufacturing sector) with that for other firm categories.
10. The numbers in brackets show the number of dummy variables included in the equations estimated.
11. See Pearce (1990) for a detailed discussion of tenure effect on firm size–wage relationship.
12. White-collar workers include managers, administrators and clerks, engineers and physical scientists. Blue-collar workers include production workers, sales workers, technicians, supervisors and foremen, equipment maintenance, electricians, plumbers, welders, masters/crafts, apprentice and support staff.
13. The estimates are based on cash wage payment. It is possible that this underestimates the level of wages in small firms since most of the small firms are owner-managed and most of the payment is often in kind.
14. See White (1980) for a detailed discussion of this approach.
15. The estimates are obtained using the formulae $(e^{\beta} - 1)*100$, where β is the estimated coefficient of the small, medium, large and very large firms size in the first row of Table 5.2.
16. Velenchik (1997a) obtained the relative firm–size effect of medium, large and very large firms on wages to be 0.386, 0.623 and 1.056, respectively, after controlling for measured labour quality. Exact comparison is hard because of differences in the definition of firm size in these studies and ours.
17. See, for example, Evans and Leighton (1989) and Idson and Feaster (1990) for some discussion on sorting of workers across different firm sizes.
18. According to Brown and Medoff (1989), detailed controls for industry and occupation should capture a greater proportion of the variations in working conditions, because most of these variations occur across industries and occupations.
19. The sectors were divided as follows: textiles into (1) textiles and (2) garment; wood into (1) wood and cork products, and (2) furniture and fixtures; food into (1) bakery products, (2) grain mill products, (3) meat and dairy products and canned vegetables, fish, oils and fats, (4) sugar, confectionery, beverages and tobacco, and (5) miscellaneous foods.
20. Longer hours of work can give an indication of inconvenient working hours, shifts could be a measure of irregular and inconvenient hours of work and also non-permanent employees are not likely to have better terms of work than the permanent employees because they are not entitled to some benefits which are available to permanent employees.
21. We did not include all factors that control for undesirable working conditions in the wage equation because of limited information: for example, variables

indicating dangerous and unhealthy working conditions on the job and whether danger/threat is a serious problem, and variables giving commuting time among others. Additional control for these variables is likely to reduce the wage premium associated with firm size, but most unlikely to eliminate it. Brown and Medoff (1989) controlled for working conditions using most of these variables, but still they could not eliminate the size–wage differentials.

22. We used the predicted values for hourly wages in the tenure regression to avoid endogeneity problem.
23. An example of this internal labour markets in Kenya are seniority scales. Bigsten and Collier (1980), in their study on consequences of labour turnover in Kenya, indicate that seniority scales may be introduced to reduce turnover.
24. See Idson (1996) for detailed analysis on the relationship between employer size and labour tenure.
25. Of the total 767 trade disputes studied by Sandbrook (1975) in Kenya between 1969 and 1972, around 68 per cent were related to dismissals while 11 per cent were due to terms of service.
26. For more information on trade unions in Kenya see for example, Bigsten (1984), Chege (1988) and House and Rempel (1976b).
27. The implication of using profits per worker directly instead of an instrumental variable is that the coefficient of profit variable will be underestimated (see, e.g., Teal, 1995).
28. We included capital intensity to capture the firms' ability to pay, but the estimate gave an unexpected sign and in addition it had negligible effect on the firm size–wage effect, and therefore was excluded.
29. Collier and Gunning (1999) note that rents which government interventions give rise to are not volatile and hence under rent-sharing hypothesis profitability should be significant in a cross-section (explaining wage levels), but not in a panel estimate (explaining first differences).

References

Baltagi, H. B. (1995), *Econometric Analysis of Panel Data*, John Wiley, New York.

Bigsten, A. (1984), *Education and Income Distribution in Kenya*, Gower, Alderstot.

Bigstein, A. and Collier, P. (1980), 'Economic Consequences of Labour Turnover in Kenya: A Pilot Application of the National Social Security Fund Data', mimeo, Oxford.

Blackburn, M. and Neumark, D. (1992), 'Unobserved Ability, Efficiency Wages and Interindustry Wage Differentials', *Quarterly Journal of Economics* 107.

Blanchflower, D. G., Oswald, A. J. and Sanfey, P. (1996), 'Wages, Profits and Rent-sharing', *Quarterly Journal of Economics* 111.

Brown, C. and Medoff, J. (1989), 'The Employer Size–Wage Effect', *Journal of Political Economy* 97.

Chege, M. (1988), 'The State and Labour: Industrial Relations in Independent Kenya', in Coughlin *et al.*, *Industrialisation in Kenya: In Search of a Strategy,* Heinemann, Nairobi and James Currey, London.

Collier, P. and Gunning, J. W. (1999), 'Explaining African Economic Performance', *Journal of Economic Literature* 37.

Departments of Economics, University of Göteborg and University of Nairobi (1994), *Limitations and Rewards in Kenya's Manufacturing Sector: A Study of Enterprise Development*, World Bank, Washington, DC.

Duncan, G. and Holmlund, B. (1983), 'Was Adam Smith Right After All? Another Test of the Theory of Compensating Wage Differentials', *Journal of Labour Economics* 1(4).

Evans, D. S. and Leighton, L. S. (1989), 'Why do Smaller Firms Pay Less?', *Journal of Human Resources* 24.

Freeman, R. B. and Medoff J. L. (1981), 'The Impact of the Percentage Organised on Union and Non Union Wages', *Review of Economics and Statistics* 63.

Gerlach, K. and Schmidt, E. M. (1990), 'Firm Size and Wages; *Labour* 4.

House, W. J. and Rempel, H. (1976a), 'The Determination of and Changes in the Structure of Wages and Employment in the Manufacturing Sector in the Kenya Economy 1967–1972', *Journal of Development Economics* 3.

House, W. J. and Rempel, H. (1976b), 'The Impact of Unionisation on Negotiated Wages in the Manufacturing Sector in Kenya', *Oxford Bulletin of Economics and Statistics* 38.

Hsiao, C. (1986), *Analysis of Panel Data*, Cambridge University Press, New York.

Idson, T. L. (1996), 'Employer Size and Labour Turnover', in *Research in Labour Economics* 15.

Idson, T. L. and Feaster, D. J. (1990), 'A Selectivity Model of Employer Size–Wage Differentials', *Journal of Labour Economics* 8.

Ikiara, G. K. and Ndung'u, N. S. (1997), 'Employment and Labour Market During Adjustment: The Case of Kenya', ILO, Geneva.

Knight, J. (1998), 'Labour Market Policies and Outcomes in Post-Independence Zimbabwe: Lessens for Southern Africa', in Lennart Petersson (ed.), *Post-Apartheid Southern Africa: Economic Challenges and Policies for the Future*, London, Routledge, pp. 201–6.

Knight, J. B. and Sabot, R. H. (1990), *Education, Productivity and Inequality: The East African Natural Experiment*, Oxford University Press, New York.

Kostiuk, P. (1990), 'Compensating Differential for Shift Workers', *Journal of Political Economy* 98(5).

Krueger, A. B. and Summers, L. H. (1988), 'Efficiency Wages and Inter-industry Wage Structure', *Econometrica* 56.

Lazear, E. P. (1981), 'Agency, Earnings Profiles, Productivity, and Hours Restrictions', *American Economic Review* 71.

Lester, R. (1967), 'Pay Differential by Size of Establishment', *Industrial Relations* 7.

Masters, S. H. (1969), 'An Inter-industry Analysis of Wages and Plant Size', *Review of Economics and Statistics* 51.

McCormick, D. (1993), 'Risk and Growth, the Dilemma of Nairobi's Small Scale Manufacturers', Discussion Paper No. 291, Institute of Development Studies, University of Nairobi.

McCormick, D., Kinyanjui, M. N. and Ongile, G. (1997), 'Growth and Barriers to Growth Among Nairobi's Small and Medium-Sized Garment Producers', *World Development* 25(7).

Mcnabb, R. (1989), 'Compensating Wage Differentials: Some Evidence from Britain', *Oxford Economic Papers* 41.

Mellow, W. (1982), 'Employee Size and Wages', *Review of Economics and Statistics* 64.

Neitzert, M. C. (1992), 'A Microeconomic Analysis of the Urban Market in Kenya', PhD Thesis, University of Toronto.

Oi, W. Y. (1990), 'Employment Relations in Dual Labour Market ("It's Nice Work If You Can Get It")', *Journal of Labour Economics* 8.

Ongile, G. and McCormick, D. (1996), 'Barriers to Small Firm Growth: Evidence from Nairobi's Garment Industry', in Dorothy McCormick and Paul O. Pedersen (eds), *Small Enterprises: Flexibility and Networking in an African Context*', Longhorn Kenya, Nairobi.

Pearce, J. F. (1990), 'Tenure, Unions and the Relationship Between Employer Size and Wages', *Journal of Labour Economics* 8.

Sandbrook, R. (1975), *Proletarian and African Capitalism: The Kenyan Case*, Cambridge University Press, Cambridge.
Stiglitz, J. E. (1974), 'Alternative Theories of Wage Determination and Unemployment in LDCs: The Labour Turnover Model', *Quarterly Journal of Economics* 88.
Teal, F. (1995), 'The Size and Sources of Economic Rents in a Developing Country Manufacturing Labour Market', *Economic Journal* 106.
Velenchik, A. D. (1997a), 'Government Intervention, Efficiency Wages, and the Employer Size Wage Effect in Zimbabwe', *Journal of Development Economics* 53(2).
Velenchik, A. D. (1997b), 'Market Power, Firm Performance and Real Wage Growth in Zimbabwean Manufacturing', *World Development* 25(5).
White, H. (1980), 'A Heteroscedasticity-Consistent Covariance Matrix Estimator and a Direct Test for Heteroscedasticity', *Econometrica* 48(4).

6
Financial Constraints on Kenyan Manufacturing

Anders Isaksson and Clas Wihlborg[1]

6.1 Introduction

In 1993, no fewer than 80 per cent of the managers participating in the RPED survey in Kenya mentioned lack or costs of financing as a moderate or major obstacle to the operations and expansion of their firms. It can be expected that many managers perceive access to or costs of financing as constraints. The issue of concern is whether financing is forthcoming for the projects that provide an acceptable rate of return. The RPED survey data on various forms of financing allow us to analyse in more detail firms' financial conditions, and the availability and choice of different forms of financing. The major purpose of our previous studies of the survey responses was to discover whether access to financing was a severe constraint on manufacturing activity in Kenya during the years 1993 and 1994 (Isaksson and Wihlborg, 1994, 1995). In this chapter we extend the analysis to include results of the 1995 survey. With three years of data we can obtain greater reliability and take changing macroeconomic conditions into account.

Availability of financing may be considered a constraint on economic activity for different reasons. First, it is not unusual for governments to impose interest rate ceilings with the result that demand for credit must be rationed. This reason for 'financial repression' is less common now than it was in the 1970s and the 1980s, because interest rate determination is left to the market in an increasing number of countries. Second, the influence of interest groups through political interference in banking or the direct influence of groups closely associated with particular banks may lead to a credit allocation process whereby many firms with viable projects are shut out while other firms are able to get access to credit on favourable terms. Third, the legal institutional framework for enforcement of financial contracts may be weak, with the consequence that lenders must either rely on self-enforcement mechanisms such as the borrower's reputation or not give loans.

Even in the most advanced industrialised countries there are 'inefficiencies' in financial markets as compared to a textbook, competitive, full information model. In particular, the asymmetry of information about project return and about the honesty of the borrower creates the need for a variety of institutional and contractual arrangements with the purpose of minimising the information problem and securing contract enforcement. Many institutional and contractual arrangements are developed in the marketplace but some credit rationing is inevitable and may be considered an efficient response to remaining information asymmetries. We want to speak about the financial sector as a constraint on economic activity only if the legal institutional framework hinders enforcement of potential voluntary contracts or if incentives for entering such contracts are lacking because of, for example, political influences on the credit allocation process. We return to these issues below.

In our previous studies of financing as a constraint on manufacturing activity in 1993 and 1994 we relied both on descriptive data on various forms of financing and on more formal statistical analysis of the association between financing and firm characteristics capturing the degree of asymmetry of information and the degree of self-enforcement of contracts. In this chapter we extend the analysis with data for one more year but we limit the analysis of all three years to a study of descriptive data on financing. We proceed as follows: in section 6.2 an overview of macroeconomic developments and conditions in the financial sector is provided. Section 6.3 contains a summary of results of our previous studies based on data for the years 1992 and 1993. Financial data for Kenyan manufacturing firms for each of the years 1992–94 are presented and discussed in section 6.4. We focus on trade credits and informal loans in section 6.5, asking whether these forms of financing substitute for many firms limited access to formal loans in banks and other financial institutions. Section 6.6 summarises the results.

6.2 Macroeconomic development and financial sector performance

6.2.1 Macroeconomic development, 1992–95[2]

Since the survey period is 1993–95, our overview of the macroeconomic development covers only 1992–95. We describe developments that directly or indirectly may be relevant for conditions in financial markets.

The Kenyan economy improved considerably over the four years. While real GDP growth was barely positive in 1992 and 1993, and even negative in per capita terms, there was a strong recovery in 1994 and 1995, as shown in Table 6.1.

From 1992 to 1995, the financial sector growth rate was 6–7 per cent annually. Much of this growth can be attributed to the completion of the internal financial liberalisation programme in 1991. During the survey period interest rates were market-determined and the number of financial

Table 6.1 Macroeconomic indicators, 1992–95

	1992	*1993*	*1994*	*1995*
Real GDP growth	0.5	0.2	3.0	4.8
Real GDP growth, per capita	−2.3	−2.7	0.1	1.9
Manufacturing sector growth	1.2	1.8	1.9	3.9
Financial sector growth	6.9	7.2	6.1	6.9
Domestic credit, % of GDP	61.1	53.6	60.3	63.9
Domestic credit to public sector, % of GDP	17.3	17.2	22.4	19.4
Domestic credit to private sector, % of GDP	43.8	36.4	37.9	44.5
Credit to private sector/total credit	68.0	69.0	62.0	66.4
Money supply growth, M1	50.0	23.0	11.0	9.0
Money supply growth, M2	35.0	28.0	30.4	18.6
M1/M2	60.0	49.0	41.0	38.0
Treasury Bill rate, 91 days	16.6	48.1	24.0	19.0
Real Treasury Bill rate, 91 days	−12.9	2.3	−5.0	17.4
Savings deposit rate, max.	25.7	25.6	16.9	11.9
Lending rate, max.	30.0	31.6	28.3	27.2
Spread between lending and savings rates	4.3	6.0	11.4	15.3
Inflation	29.5	45.8	29.0	1.6
Exchange rate, Ksh/US$	36.2	68.2	44.8	55.9

Note that the fiscal year runs from 1 July to 30 June.
Source: *Economic Survey* (various issues); Isaksson and Kulundu (1998).

instruments available to the public increased. External financial liberalisation was ongoing as well during this period and access to foreign exchange improved. International capital flows were liberalised. Nevertheless, the Kenyan financial market remained relatively segmented from world markets.

Total domestic credit as a share of GDP remained constant at around 61–2 per cent between 1992 and 1995, except for a significant drop in 1993. This drop was entirely attributed to a decrease in credit available to the private sector. In 1994 the share of credit to the private sector fell further, from 69 per cent in 1993 to 62 per cent. This could be a sign of crowding out. The private sector's share of domestic credit picked up again in 1995.

In 1993, the manufacturing sector received approximately 19 per cent of the banks' and 14 per cent of the Non-Bank Financial Institutions' (NBFIs) credit. This corresponds largely to the manufacturing sector's share of GDP. The average growth rates for the manufacturing sector and for the overall economy were approximately the same, but credit directed towards the private sector declined in 1993 and 1994.

After the election-induced monetary expansion of 1992, the central bank aimed at reducing liquidity. There was a sharp decrease in the growth of narrow money (M1) relative to the broader measure (M2). In 1992 the ratio of M1 to M2 was 68 per cent, while in 1995 the ratio had dropped to

38 per cent (Table 6.1). The expansion of the money supply in 1992 was the main reason for the high inflation and the build-up of inflationary expectations between 1992 and 1995. The high inflation led to high nominal interest rates. The public's expectations about future inflation remained high during the monetary contraction after 1992. This combination of factors explains the high real interest rates in 1995 shown in the table. In 1994 both the savings interest rate and the nominal Treasury Bill rate fell, but the lending rate fell at a slower pace. As a result the interest rate spread between the lending and savings rates doubled in 1995. It can be expected that these developments led to a reduction in investments although the financial liberalisation may have dampened the negative effect on investments.

The fiscal situation was weak in the financial years 1992/93 and 1993/94, with an overall budget deficit amounting to 4 per cent and 7 per cent of GDP, respectively. In the following two years the overall fiscal deficit decreased to about 1 per cent of GDP. This improvement was mainly due to a successful reduction in expenditures but also to increasing external grants. Financing of the government deficit shifted towards domestic borrowing in the 1990s, which partly explains the increasing interest rates. Increased domestic public sector borrowing usually means some degree of crowding-out of the private sector in financial markets. In the Kenyan case the rising real domestic interest rates in the 1990s is a reflection of this crowding out.

For firms doing business internationally the large changes of the exchange rates shown in Table 6.1 were a serious nuisance, because almost all real capital and most raw materials are imported.

6.2.2 The financial sector

By sub-Saharan African standards, Kenya has a well-developed financial system. The financial sector has developed further since the implementation of the financial sector reform initiated in 1989. The reform was assisted by the IMF and the World Bank, although the government had already started implementing some reforms in 1988. There were reforms of both policy instruments and institutions. The former reforms included interest-rate liberalisation, development of money and capital markets, improvement of the efficiency of financial intermediation, development of more flexible monetary-policy instruments, removal of credit ceilings and reduction in the government's reliance on domestic bank borrowing, coupled with a reduction of the budget deficit. Institutional reforms were aimed at creating a regulatory framework and ensuring prudential regulation and supervision of the financial system. Financial institutions were also restructured and privatised.

The first donor-supported financial sector reform lasted throughout 1991, but it was soon followed by additional reforms initiated by the government of Kenya itself. They included measures to ensure current and capital account convertibility. The Central Bank increasingly undertook open market operations, improved reserve money management and the regulation of the banking system.

By 1993, the number of operational commercial banks and NBFIs were 30 and 51, respectively. There were five building societies. The restructuring of some unsustainable financial institutions throughout 1995 led to a decrease in the number of NBFIs, due to conversion into commercial banks, merging with parental commercial banks, or closure. In 1995, only 39 NBFIs remained, while the number of commercial banks increased to 41. The number of building societies decreased to four.

Only four large banks dominate the banking sector in Kenya. In 1997 these banks together controlled 50 per cent of bank assets. The largest bank in the system was the state-owned Kenya Commercial Bank, with 17 per cent of bank assets. It was followed closely by the foreign-owned Barclays Bank, with 16 per cent of bank assets and 15 per cent of bank deposits. The group of large banks also included the state-owned National Bank of Kenya and the foreign-owned Standard Chartered Bank, with 8 per cent each of bank assets (IMF, 1998).

By the end of 1995, non-performing loans of the 74 institutions supervised by the Bank Supervision Department amounted to 17.6 per cent of total loans.[3] Clearly many banks had negative net worth and were in need of restructuring.

The financial system is modelled on the British system. One implication of this is that commercial banks are not allowed to hold equity for purpose of control and as a form of lending to firms.

The Nairobi Stock Exchange (NSE) was established in 1954 and is overseen by the Capital Markets Authority (CMA), the supervisory body in charge of capital markets development in Kenya. Trading at NSE increased rapidly, although few firms were listed in the early 1990s. In 1995, there were 56 firms were listed at the NSE. The regulation of foreign ownership of shares in local firms was removed in 1995. The main reason for this was to attract foreign capital. Issuance of commercial and government papers and Treasury Bills was also introduced in 1995.

Informal financial markets are often viewed as important, especially for 'quality loans' (see, for example, World Bank, 1993). Based on the data presented below the importance of organised informal markets in Kenya must be doubted. To the extent that informal finance is important, it is the funding by relatives and friends that count. We will call these suppliers of informal finance 'angels', and distinguish them from more 'institutionalised' suppliers, such as money-lenders and saving clubs.

An additional aspect of the financial sector is the legal institutional framework as mentioned in the Introduction. Although there was financial sector reform in Kenya in the period under study, the legal institutions supporting financial contracting remained weak. As discussed in Eliasson,

Rybczynski and Wihlborg (1993) the state must be the 'enforcer of last resort' of contracts unless Mafia-like contract enforcement is considered acceptable. Furthermore, property rights to assets at stake in contracts must be well defined and transferable in case of non-performance. For example, much lending in most countries requires collateral with the implications that borrowers promise to give up property in case of non-payment. Unless ownership of property can be proved and property rights can be transferred to the lender, such collateral is without value. The same is true if courts do not enforce transfer of the collateral assets or if bankruptcy proceedings take long and outcomes are uncertain. In the absence of transferable assets as collateral for financial contracts, and in the absence of a court-system threatening some kind of repercussion in case of non-payment, a lender must rely on self-enforcement mechanisms. In the case of Kenya it is reasonable to expect that lenders must rely on self-enforcement to a high degree.

Unless complete honesty can be taken for granted, self-enforcement depends on the borrower having something at stake in the fulfilment of a contract. An individual's or a firm's reputation is an example of an 'asset' that can be staked. The fear of loss of reputation works as an enforcement mechanism if business or social relations depend on contract fulfilment. Self-enforcement is an imperfect enforcement mechanism for many types of transaction. Thus reliance on it would exclude many potentially viable transactions.

Studies indicate that the Kenyan court system does not enjoy the confidence of the private sector.[4] One reason is that there is a perception that judges lack autonomy relative to political influences. Executive, ministerial and lower level administration enjoy a high level of discretion in commercial matters. For example, we were told in interviews that local authorities can determine that borrowers need not give up assets given as collateral even when the borrower has been found in default. It is especially land-use and transfers of land ownership that is controlled by such local authorities (Land Control Boards). The consent of these authorities is required for transfers and sales, and requirements vary among Land Control Boards across the country.

Another problem is that the court system is slow in resolving disputes. Bankruptcy and liquidation are extremely slow processes, taking 4–10 years to complete. The value of collateral against a loan can be seriously questioned under these circumstances.

The value of land as collateral is also influenced by property legislation and the completeness of registration of and title to property. There are serious deficiencies both in registration and in titles. Large areas are government owned and leased for 99 years. Other land areas are controlled by groups of individuals. Only 3.6 million hectares were individual parcels of land in 1983.

What can be expected from firm-level financial data during the 1992–95 period following financial sector reform? In spite of the weak legal institutions, we expect that the many reforms should have led to improved access to formal credit for many firms. We expect a move away from the use of informal financing among the same firms. The use of trade credit as a substitute for bank loans should also decrease. Nevertheless we expect that large groups of firms have very limited access to formal financial markets and perhaps also to trade credits. These expected results might be blurred by macroeconomic developments such as the tendencies of crowding-out mentioned above. At times Treasury Bills offered an attractive investment opportunity for non-financial firms as well.

6.3 Results from analysis of 1993 and 1994 data

Analysis of data for 1993 revealed that access to formal loans seemed to be restricted for nearly all groups of manufacturing firms. Small and 'micro' firms in particular had little access to formal loans. Entrepreneurs faced grave difficulties in raising start-up funds unless they had a wealthy friend or relative serving as an angel. The lack of enforcement of collateral seemed to be an important limiting factor for an individual seeking, for example, start-up funds. The very high collateral/loan ratio for all firms indicated that even large firms had constrained access to formal loans. Informal financial institutions did not seem to play a large role. However, angels (friends and relatives) played an important role for financing among Africans for start-up financing, and, in particular, among Kenyans of non-African and non-Asian origin in Mombasa.

We did not find that trade credits were important in terms of amounts, but the data were weak on this point. A sizeable proportion of all firms uses these credits, however. It seemed that small Asian-owned firms obtained credits from suppliers contributing to the financing of operations. Small, often informal, African firms obtained some funds (net) from customers making advance payment. Multinationals' subsidiaries obtained a very substantial trade credit financing, most likely from abroad. These firms did not seem to supply credits to other firms correspondingly.

There was clearly a very large proportion of small and informal firms working entirely in a cash economy. Self-enforcement mechanisms, as well as legal enforcement of contracts, did not seem to function very well for these firms. The same group of firms had to rely very heavily on their own savings for start-up financing and retained earnings for financing of additional investments.

The interview study did not give any indication of any perverse or discriminatory behaviour on the part of the banks. It was known, however, that funds in the banking system had been diverted to some extent towards the financing of parastatals.

Formal analysis using probit and OLS regression methods was employed in order to discover whether access to formal loans hindered business operations and reduced investments. It was found that limited liability firms in general had greater access to loans from formal institutions than other firms did. The reason is probably that such firms are well-defined legal entities. Their assets are therefore also more easily identifiable and contracts more enforceable.

Unfortunately, the first year's study was performed on data collected in a severe recession year with relatively little investment activity. Nevertheless, we found indications that access to formal loans had constrained investment activity. This conclusion was based on similarities of explanatory factors for the likelihood that a firm obtained formal loans and the likelihood that a firm invested in 1993. It was also found that the rate of profits influenced both the share of loans relative to assets, and the rate of investment positively among firms that borrowed and invested.

The analysis of 1994 data confirmed and strengthened most of the mentioned results. The higher quality of the data (in particular for trade credits and advance payment) had important effects on the results. Trade credits and advance payments as sources and uses of funds played a much more prominent role in the 1994 data for exactly the same groups of firms as in 1993 (with the exception of a few firms which no longer existed).

We concluded again that neither informal loans nor trade credits received substituted effectively for access to formal loans. Although angels in the form of friends and relatives were important sources of financing for many small firms, other informal sources of finance did not seem to play a major role. Firms lacking access to formal loans generally lacked access to trade credits as well. However, they had to offer trade credits to their customers in many cases. The only financial relief some micro and small firms seemed able to obtain was advance payments from some customers. Access to trade credits, and their volume relative to assets, were strongly and positively associated with the value of collateral for formal loans, as well as with access to formal loans. It was also found that firms lacking formal loans more frequently pay in advance. These results confirm the hypothesis that trade credits and advance payments do not substitute for access to formal loans as sources of financing.

The share of firms receiving formal loans, including overdrafts, was nearly 70 per cent, but the figure ranged from 23 per cent of micro firms to 100 per cent of macro firms. These loans represented about 1 per cent of the micro firms' assets, 7–8 per cent of small firms' assets, and 22–4 per cent of the largest firms' assets.[5] There were no clear differences related to ethnicity.

The formal probit and OLS regression analyses were conducted with a limited, pre-determined set of explanatory characteristics that were intended to capture information availability and enforcement. The ability to offer collateral and the value of collateral given were clearly explanatory factors for different sources of financing, as well as for investments. Current profits played a lesser role in 1994 than in 1993, however.

In addition to the strong results for collateral some other risk, information and enforcement variables had explanatory value for sources of financing: for instance, the strength of local competition helped explain the availability of financing. Variables expressing availability of information and the strength of self-enforcement mechanisms explained primarily access to trade credits and their volume relative to assets. Business condition variables played a lesser role, although correlation with the collateral variable obscured results. Regulatory conditions had little explanatory value.

6.4 The structure of financing and formal borrowing, 1993–95

In this section we present data for the structure of financing of manufacturing firms in Kenya 1993–95. An aggregate balance sheet has been calculated for each year and for the firms in different size groups. The percentage of all firms relying on different types of financing and formal loans in particular are also presented. We do not disaggregate the firms by sector and by formal versus informal firms as in the previous studies. The information obtained by these kinds of disaggregation is not going to be enhanced relative to our previous studies by the addition of data for 1995. We distinguish between firms owned by different ethnic groups, however, since changes in the relative economic activity may have political significance in Kenya. Thus, we distinguish among Asian-(owned), African-(owned), and other firms in the tables below.

All the data presented here refer to 'surviving' firms, meaning that the firms responded to the questionnaire in each of the three years. In some cases of 'drop-outs', when possible, a new firm in the same category was substituted for the 'drop-out'. Thus, the tables below could be presented for all firms in the sample as well. The differences in results for 'surviving' and 'all' firms are negligible. By presenting the data for surviving firms we accomplish comparability across the years, while the data are also reasonably representative for the manufacturing sector as a whole.

In order for the data to be representative for the manufacturing sector in Kenya as a whole a weighting scheme is applied on answers from each firm. The scheme is described in Chapter 3. However, there is a potential source of bias when using employment-based weights for analysing financial data for assets and liabilities. This bias could have been severe if all sectors had been included but as long as the analysis is limited to regular manufacturing sectors the bias is likely to be small.

Tables 6.2a and 6.2b, respectively, show the assets and liabilities of firms, focusing on financial items in order to reveal sources of finance as well as firms' contributions to other firms financing. 'Equity and retained earnings' is

Table 6.2a Assets by size and ethnicity, percentage of total assets 1993–95 ($N = 193$)

		Micro			*Small*			*Medium*			*Large*			*Macro*		
		Asian	*African*	*Other*	*Asian*	*African*	*Other*	*Asian*	*African*	*Other*	*Asian*	*African*	*Other*	*Asian*	*African*	*Other*
Real assets	1993	99.85	98.10	–	97.62	99.50	–	93.58	98.17	–	96.02	91.00	79.85	96.65	94.57	94.00
	1994	92.28	96.38	–	94.36	96.67	–	93.00	94.51	–	93.21	84.60	92.28	95.41	–	77.12
	1995	92.35	94.92	–	94.41	95.66	–	88.69	89.43	–	92.98	83.76	84.48	99.30	–	–
Informal lending	1993	0.00	0.51	–	0.10	0.06	–	0.05	0.29	–	0.11	0.00	0.00	0.00	0.00	0.00
	1994	7.58	1.46	–	0.52	0.62	–	0.04	0.03	–	0.05	2.19	0.00	0.01	–	0.00
	1995	6.83	2.53	–	0.08	0.54	–	0.11	0.31	–	0.23	0.00	0.05	0.00	–	–
Trade credit (receivables)	1993	0.15	1.42	–	1.94	0.34	–	6.10	1.54	–	3.87	8.57	10.08	3.35	5.43	6.00
	1994	0.13	3.32	–	4.78	2.60	–	6.69	5.46	–	6.34	13.21	7.42	4.58	–	22.88
	1995	0.80	2.49	–	5.01	3.06	–	11.00	9.24	–	6.60	15.15	15.35	0.45	–	–
Advance payment given	1993	0.00	0.00	–	0.00	0.10	–	0.26	0.00	–	0.00	0.42	10.08	0.00	0.00	0.00
	1994	0.00	0.15	–	0.32	0.11	–	0.27	0.00	–	0.40	0.00	0.30	0.00	–	0.00
	1995	0.02	0.05	–	0.51	0.75	–	0.21	1.01	–	0.18	1.08	0.11	0.26	–	–

Note: Size groups are defined as: Micro = 1–5 employees; Small = 6–20 employees; Medium = 21–75; Large = 76–500; Macro = 501+.
N stands for number of observations. Weights have been used in all calculations.

Table 6.2b Liabilities by size and ethnicity, percentage of total assets 1993–95 ($N = 193$)

		Micro			*Small*			*Medium*			*Large*			*Macro*		
		Asian	*African*	*Other*	*Asian*	*African*	*Other*	*Asian*	*African*	*Other*	*Asian*	*African*	*Other*	*Asian*	*African*	*Other*
Loans by angels	1993	0.01	1.49	–	0.17	0.31	–	0.04	0.08	–	0.22	0.00	0.00	0.26	0.00	0.00
	1994	0.00	0.35	–	0.24	1.74	–	0.09	0.00	–	0.40	0.00	0.00	1.12	–	0.00
	1995	0.00	1.69	–	0.14	1.34	–	0.20	0.00	–	0.25	0.00	0.00	0.00	–	–
Other informal loans	1993	0.00	0.00	–	0.00	0.00	–	0.00	0.00	–	0.00	0.00	0.00	0.00	0.00	0.00
	1994	0.00	0.64	–	0.00	0.37	–	0.00	0.00	–	0.00	0.66	0.00	0.00	–	0.00
	1995	0.00	0.06	–	0.00	0.00	–	0.00	0.00	–	0.00	0.00	0.00	0.00	–	–
Overdraft	1993	0.31	0.06	–	4.75	6.51	–	4.24	1.75	–	5.72	4.49	0.00	10.56	2.65	2.46
	1994	0.13	0.04	–	3.43	2.82	–	4.14	1.07	–	4.60	1.86	1.63	4.94	–	9.31
	1995	0.45	0.40	–	5.45	0.46	–	6.89	5.72	–	6.38	2.73	4.02	9.29	–	–
Formal loans	1993	0.25	0.03	–	4.02	1.03	–	2.52	7.52	–	5.60	5.08	0.00	1.55	2.41	16.37
	1994	0.00	0.05	–	0.65	0.05	–	2.78	2.95	–	2.40	3.84	0.00	7.94	–	8.58
	1995	41.23	0.68	–	0.93	0.02	–	2.29	13.90	–	5.10	2.78	1.30	5.08	–	–
Trade credit (payables)	1993	0.00	0.23	–	1.57	0.97	–	3.69	0.73	–	1.81	6.10	13.74	2.58	5.09	11.46
	1994	0.02	0.21	–	3.92	0.62	–	4.49	4.37	–	3.96	16.03	4.45	0.72	–	12.36
	1995	0.34	0.69	–	3.11	0.55	–	5.49	4.99	–	5.13	19.82	9.77	2.65	–	–
Advance payments received	1993	0.00	3.29	–	0.08	0.55	–	0.07	0.85	–	1.25	0.00	0.00	0.00	0.00	0.55
	1994	0.00	1.96	–	0.12	0.48	–	0.25	0.03	–	0.77	0.76	0.00	0.00	–	0.00
	1995	0.03	3.70	–	0.56	3.34	–	0.45	0.14	–	0.22	0.93	0.65	0.00	–	–
Equity and retained earnings	1993	99.43	95.98	–	89.40	90.63	–	89.44	89.05	–	85.40	84.32	86.26	85.06	89.85	69.17
	1994	99.85	96.75	–	91.63	93.92	–	88.25	91.58	–	87.88	76.85	84.27	85.27	–	69.75
	1995	57.95	92.79	–	89.81	94.28	–	84.67	75.25	–	82.93	73.73	89.02	82.99	–	–

Note: Size groups are defined as for Table 6.2a. *N* stands for number of observations. Weights have been used in all calculations.

a 'plug-figure' obtained by deducting survey data for the other types of financing from total assets. The latter variable is calculated from survey data.[6]

Turning to Tables 6.2a and 6.2b, equity and retained earnings represent more than 85 per cent of the liabilities with few expectations. Only macro firms owned neither by Africans nor Asians rely on non-owner financing to a degree of around 30 per cent in each year. These firms are primarily multinationals' subsidiaries.

Large and macro firms with Asian and African owners reach around 25 per cent of non-owner financing in the last two years.

There seems to be an anomaly for Asian micro firms in 1995. After having nearly no access to formal loans in 1993 and 1994 these firms seem to obtain 40 per cent formal debt in 1995. We are concerned that this figure may be the result of the small number of observations for Asian micro firms, since the majority of these very small firms are informal African-owned firms. We return to this issue below when data on the share of firms with access to formal loans are discussed.

It can also be seen that in 1994 small Asian firms increased their offering of trade credits substantially. Asian micro firms were also involved in informal lending to a high degree in 1993 and 1994. This instability in the balance sheets for Asian micro and small firms is another concern that we return to below.

In all size groups there is little reliance on angels and other informal loans. Trade credits as a source of financing jumped substantially between 1993 and 1994. We suspect that the surveys underestimated this source during the first year. Large firms rely most heavily on trade credits. Large African firms stand out with a trade credits share of around 15 per cent in 1994 and 1995. All Asian firms except the micro ones seem to have relatively good access to trade credits. Advance payments received appear substantial only for African micro firms. While large African firms seemed to rely on trade credits for financing, micro and small African firms issue more trade credits than they receive. Trade credits and advance payments are discussed in more detail below.

The most striking results in Tables 6.2a and 6.2b is the low share of debt financing across the board among manufacturing firms. Even the two largest groups of firms are between 75 and 85 per cent equity financed disregarding the non-African, non-Asian firms.[7] For Asian as well as African firms the equity share has been declining during the period, but there are too few time periods in the data to talk about a tendency. The general conclusion remains that the banking system seems to 'underproduce' loans.

Another way to study the access to loans is to analyse the share of firms in various groups, which have access to the formal debt market, including formal loans from financial institutions and overdraft. Tables 6.3, 6.4, and 6.5 show the percentage of firms that have been able to borrow at all in found financial institutions. Table 6.3 includes both regular loans and overdrafts, while Tables 6.4 and 6.5 show the results for either regular loans or overdrafts. The firms are divided by size groups and ethnic groups.

Table 6.3 shows that nearly 100 per cent of the largest firms, and the great majority of medium and large firms, are able to obtain formal loans. Thus, the low average debt share of these firms in the previous table does not reflect lack of borrowing by a share of them.

The second feature of Table 6.3 is that micro and small firms have very limited access to the formal loan market. The very smallest firms (micro)

Table 6.3 Percentage of firms with formal loans including overdraft, 1993–95

	1993		*1994*		*1995*	
	%	N	%	N	%	N
Micro	5.40	45	10.30	41	11.90	43
Small	38.00	35	28.10	44	10.40	41
Medium	99.10	46	80.90	40	91.40	44
Large	81.50	33	62.70	39	86.60	34
Macro	100.00	4	100.00	4	94.50	4
Formal	79.60	117	62.40	122	67.20	120
Informal	18.00	46	15.60	46	12.40	46
African	19.60	72	16.80	72	11.30	72
Asian	46.10	80	38.40	83	89.30	81
Other	6.20	11	4.80	13	3.30	3
All	20.00	163	17.20	168	14.30	166

Note: *N* stands for number of observations. Weights have been used in all calculations.

Table 6.4 Percentage of firms with formal loans excluding overdraft, 1993–95

	1993		*1994*		*1995*	
	%	N	%	N	%	N
Micro	2.40	45	6.20	41	6.10	43
Small	20.40	35	5.60	44	0.40	41
Medium	5.50	46	3.30	40	1.20	44
Large	9.50	33	8.30	39	6.00	34
Macro	0.00	4	0.00	4	0.00	4
Formal	14.50	117	7.80	122	6.40	120
Informal	8.00	46	5.90	46	4.40	46
African	8.50	72	6.40	72	2.60	72
Asian	7.90	80	0.60	83	51.00	81
Other	0.00	11	0.60	13	0.00	3
All	8.20	163	6.00	168	4.50	166

Note: *N* stands for number of observations. Weights have been used in all calculations.

Table 6.5 Percentage of firms with an overdraft facility, 1993–95

	1993		*1994*		*1995*	
	%	N	%	N	%	N
Micro	3.20	44	4.10	41	6.10	42
Small	28.10	35	23.10	44	10.50	41
Medium	97.70	49	85.90	40	96.70	45
Large	97.20	35	96.10	40	93.70	35
Macro	100.00	5	100.00	4	100.00	4
Formal	83.00	123	73.00	123	75.10	122
Informal	13.80	45	9.70	46	8.20	45
African	15.20	71	10.60	72	9.10	70
Asian	47.40	84	47.70	84	46.30	84
Other	6.80	13	6.40	13	7.00	13
All	16.10	148	11.90	161	10.60	156

Note: *N* stands for number of observations. Weights have been used in all calculations.

have increased their access from a 4 per cent share to a 13 per cent share though being able to borrow. On the other hand, within the group of the second smallest firms (small), the share of firms obtaining loans has decreased. The figures for formal and informal firms indicate that the decline of the share of small and medium firms with loans has occurred among formal firms. An increase in access to loans seems to have happened for Asian-owned firms. This observation may explain the large share of debt in 1995 for Asian micro firms. There are only four such firms in the sample. Thus, if one or two of these firms obtained loans the figure in Table 6.2b can be explained as an anomaly.

Tables 6.4 and 6.5 demonstrate that the great majority of loans were overdrafts. Among large and macro firms overdrafts dominate completely, while for small and medium firms regular loans are as common as overdraft facilities.

The following tables showing how collateral has been used to shed light on the limited access to the loan market for many firms, and on the low share of debt in the financing of larger firms with access to the market. Table 6.6 shows that in 1993 and 1994 almost all loans required collateral. In 1995 there is a substantial decline in the share of borrowing firms that put up collateral. The decline has occurred among micro and small, informal, African and Asian firms. If this result represents a tendency, access to loans among very small firms seem to have improved. However, it must be remembered that only very few micro and small firms obtain loans.

Table 6.7 shows what was said in section 6.3 for the years 1993 and 1994; the amount of collateral that must be offered per loan is two to three times the loan value. The median ratio of collateral to formal loan value is stable

Table 6.6 Percentage of formal loans collateralised, 1993–95

	1993		*1994*		*1995*	
	%	*N*	%	*N*	%	*N*
Micro	14.90	6	97.30	3	47.00	6
Small	100.00	11	98.00	7	55.20	8
Medium	75.00	23	100.00	11	88.90	14
Large	93.20	20	92.50	15	94.40	15
Macro	100.00	3	100.00	3	100.00	2
Formal	91.20	59	80.70	36	78.30	41
Informal	87.10	4	100.00	3	47.50	4
African	87.70	20	98.20	13	64.40	14
Asian	85.80	41	86.20	21	21.50	28
Other	100.00	2	100.00	5	100.00	3
All	87.60	63	97.40	39	51.60	45

Note: *N* stands for number of observations. Weights have been used in all calculations.

Table 6.7 Ratio of collateral to formal loan value, mean and median, 1993–95

	1993			*1994*			*1995*		
	Mean	*Median*	*N*	*Mean*	*Median*	*N*	*Mean*	*Median*	*N*
Micro	0.20	1.85	6	28.95	23.00	3	4.84	0.67	6
Small	1.78	2.00	11	33.56	4.38	7	3.84	1.44	8
Medium	6.75	2.50	23	3.52	3.03	11	3.83	2.11	14
Large	3.80	2.50	20	5.33	1.74	15	7.46	4.00	15
Macro	3.65	1.83	3	3.30	2.50	3	1.68	3.71	2
Formal	3.93	2.50	59	3.47	2.42	36	4.25	2.31	41
Informal	1.39	1.50	4	31.08	36.00	3	4.91	1.67	4
African	1.43	2.04	20	30.02	3.29	13	6.50	2.60	14
Asian	4.88	2.50	41	4.64	2.50	21	0.96	2.11	28
Other	2.33	1.88	2	2.44	1.60	5	4.66	4.00	3
All	1.62	2.27	63	28.50	2.50	39	4.85	2.31	45

Note: *N* stands for number of observations. Weights have been used in all calculations.

across the years although the mean in 1994 increased to very high levels. Few non-representative firms with extremely high ratios explain this figure. Even the median ratio is very large, however, in the sense that it implies a constraint on the supply of loans. If banks require collateral of values two or three times the loan, even large firms cannot finance a large proportion of their assets with debt.

The micro firms that are mostly informal have a relatively low ratio of collateral to loan. The reason is probably that these firms have few assets

Table 6.8 Interest rates on formal loans, mean and median, 1993–95

	1993			1994			1995		
	Mean	*Median*	*N*	*Mean*	*Median*	*N*	*Mean*	*Median*	*N*
Micro	23.65	8.25	6	–	–	–	16.97	20.00	6
Small	17.41	19.00	11	–	–	–	23.10	23.50	8
Medium	22.64	19.00	27	–	–	–	22.97	23.50	13
Large	18.43	19.00	22	–	–	–	22.62	23.00	17
Macro	17.52	20.00	3	–	–	–	14.00	17.00	2
Formal	17.13	19.00	64	–	–	–	21.62	23.00	42
Informal	19.78	19.00	5	–	–	–	16.91	20.00	4
African	19.61	18.00	22	–	–	–	16.08	23.00	15
Asian	19.10	19.25	45	–	–	–	20.33	22.50	28
Other	15.65	17.50	2	–	–	–	18.86	23.00	3
All	19.58	19.00	69	–	–	–	17.35	23.00	46

Note: *N* stands for number of observations. Weights have been used in all calculations.

that can be used as collateral. The result is that most of them are shut out from the loan market.

Table 6.8 concludes this section with interest rate data for formal loans. Unfortunately, questions about interest rates were not directly asked in the second survey and the indirect measure turned out not to be very useful. That year's (1994) responses are therefore excluded. There is some interesting information in Table 6.8. Even if we disregard the low 1993 rates for micro firms, the general level of interest rates has increased as well, as noted in section 6.2. There is one exception: the median rate for the few macro firms has fallen. The pattern across types of firms shows that interest rates fall with size.

Micro firms break the pattern but, as in 1993, the answers in 1995 may be unreliable. The difference between Asian and African firms is too small to be statistically significant.

6.5 Informal loans, trade credits and advance payments

There was little evidence of informal loans playing a significant role in the financing of Kenyan firms in 1993 and 1994. Table 6.2b above indicated that the quantitative importance of informal loans remained low in 1995. We have found already that in 1994 a substantial number of firms had obtained loans from relatives and friends (angels), however. The data for informal loans could be underestimated because managers in interviews may be reluctant to admit to using this form of financing. The managers in the RPED survey could be expected to become more open after repeated interviews. Using Tables 6.9 and 6.10 we ask whether informal loans increased in importance in the surveys from 1993 to 1995.

Table 6.9 shows that there was an increase from 1993 to 1994 in the percentage of firms responding that they had borrowed informally, either from so-called angels or from more institutionalised informal lenders. The table also shows that there was a decrease from 1994 to 1995. Thus, if increased truthfulness is a factor it would seem that 'maximum' truthfulness

Table 6.9 Percentage of firms with informal loans including angels, 1993–95

	1993		*1994*		*1995*	
	%	*N*	%	*N*	%	*N*
Micro	15.70	45	25.00	41	14.50	43
Small	16.10	35	33.70	44	23.80	42
Medium	1.50	48	7.20	40	0.40	45
Large	3.10	35	15.90	40	2.60	35
Macro	21.70	5	15.10	4	0.00	4
Formal	11.40	122	11.20	123	3.10	123
Informal	15.30	46	28.10	46	17.00	46
African	12.70	72	26.50	72	17.60	72
Asian	5.10	83	5.50	84	1.20	84
Other	93.00	13	91.40	13	0.60	13
All	15.10	168	27.50	169	16.50	169

Note: *N* stands for number of observations. Weights have been used in all calculations.

Table 6.10 Percentage of firms with informal loans excluding angels, 1993–95

	1993		*1994*		*1995*	
	%	*N*	%	*N*	%	*N*
Micro	0.00	45	6.30	41	3.30	43
Small	0.00	35	8.10	44	0.20	42
Medium	0.00	48	0.00	40	0.40	45
Large	0.00	35	5.70	40	0.00	35
Macro	0.00	5	0.00	4	0.00	4
Formal	0.00	122	4.60	123	1.30	123
Informal	0.00	46	6.90	46	2.40	46
African	0.00	72	7.30	72	2.50	72
Asian	0.00	83	0.00	84	0.00	84
Other	0.00	13	0.00	13	0.00	13
All	0.00	168	6.80	169	2.30	169

Note: *N* stands for number of observations. Weights have been used in all calculations.

was obtained in 1995. For this reason we put some faith in the responses for the last two years.

Comparing Tables 6.9 and 6.10 it is clear that angels are responsible for most of the informal loans. They are clearly important for micro and small firms, although there was a decline in the percentage of firms borrowing from this source of funds in 1995.

Most of the informal borrowing occurs among informal firms with African owners. This observation holds for both angels and more institutionalised informal sources such as moneylenders.

Both in 1993 and 1994, but not in 1995, almost all non-African, non-Asian firms responded that they had loans from angels. The study of 1994 data showed that these firms included either micro firms or the largest macro firms. It is puzzling that the largest firms rely on financing from angels, but it is possible that a manager of a subsidiary of a foreign firm made the interpretation that angels include the foreign parent. Table 6.9 shows that the informal borrowing by non-African and non-Asian firms was almost down to zero in 1995.

Table 6.10 reveals that only a few African, informal, micro-firms borrowed from money lenders or other more institutionalised informal lenders in 1995. There was a decline in this type of borrowing relative to 1994.

If informal loans did not play a very large role substituting for bank loans, it is possible that trade credits and advance payments did. Tables 6.11–14 shed some light on this issue. As noted in section 6.2, we found in the previous studies that trade credits did not effectively substitute for formal loans, but some firms were able to use trade credits to finance advance payments to suppliers.

Table 6.11 Percentage of firms receiving trade credit, 1993–95

	1993		*1994*		*1995*	
	%	*N*	%	*N*	%	*N*
Micro	2.30	44	12.50	41	16.20	43
Small	21.20	35	31.40	44	45.40	41
Medium	89.70	48	71.80	40	86.00	44
Large	36.00	35	64.80	40	83.10	34
Macro	100.00	5	34.90	4	85.60	4
Formal	34.80	122	56.80	123	64.80	120
Informal	11.40	45	18.20	45	25.10	46
African	12.20	71	19.40	72	24.00	72
Asian	19.90	83	32.70	84	39.70	81
Other	3.60	13	5.40	13	94.20	13
All	12.20	167	19.50	169	26.40	166

Note: *N* stands for number of observations. Weights have been used in all calculations.

Table 6.11 shows that most firms above the small size obtained trade credit. The firms receiving credits have increased substantially from 1993 to 1994. A higher proportion of formal firms (43.8 per cent) obtained trade credit in 1995 than the proportion of informal firms. The proportion of Asian firms receiving trade credits is twice the proportion of African firms.

More than half the African firms, and particularly those of medium size or smaller, received advance payments in 1995, as shown in Table 6.12. This proportion increased substantially from 1994 to 1995. Tables 6.13 and 6.14 for trade credits given and payments in advance reveal no similar increase in 1995. Thus, the increase in advance payments to relatively small African firms in 1995 seems to represent a real improvement in the financial situation of these firms.

Our previous conclusions, that trade credits are received largely by the same firms as those with access to bank loans, may have to be modified in 1994. The percentage of firms receiving trade credit has increased, however. In particular a higher proportion of the informal firms received trade credits (23.10 per cent) than the proportion receiving loans (13.8 per cent), while the proportions for formal firms were similar for trade credits (43.8 per cent) and loans (46.4 per cent).

Biggs, Raturi and Srivastava (1996) concluded that Asian firms in Kenya have better access to trade credits than African ones. The reason given is that the information asymmetry is lower and the enforcement possibilities better in the Asian community. Our results seem to bear this out. Our study of previous surveys showed that the pattern held for firms of similar size.

Another indication that the financial pressure on small African firms in particular was relaxed to some extent in 1995 is given in Table 6.13.

Table 6.12 Percentage of firms receiving advance payment, 1993–95

	1993		*1994*		*1995*	
	%	*N*	*%*	*N*	*%*	*N*
Micro	24.90	44	42.10	41	47.20	43
Small	66.50	35	22.10	44	66.70	41
Medium	81.20	48	23.80	40	69.00	44
Large	1.30	35	23.40	40	22.30	34
Macro	5.00	5	0.00	4	0.00	4
Formal	10.20	122	24.50	123	23.00	120
Informal	41.20	45	35.70	45	53.70	46
African	43.10	71	37.40	72	53.50	72
Asian	2.40	83	9.60	84	7.40	81
Other	1.20	13	2.90	13	91.60	13
All	40.20	167	35.30	169	52.60	166

Note: *N* stands for number of observations. Weights have been used in all calculations.

Table 6.13 Percentage of firms giving trade credit, 1993–95

	1993		*1994*		*1995*	
	%	*N*	%	*N*	%	*N*
Micro	12.60	44	52.70	41	46.40	43
Small	32.20	35	65.60	44	42.80	41
Medium	91.40	48	96.10	40	95.10	44
Large	60.30	35	78.80	40	96.70	34
Macro	95.00	5	100.00	4	8.90	4
Formal	56.90	122	81.10	123	73.20	120
Informal	21.30	45	56.70	45	46.10	46
African	19.50	71	56.80	72	45.90	72
Asian	28.00	83	45.00	84	41.30	81
Other	97.20	13	97.70	13	97.90	13
All	22.50	167	57.50	169	47.10	166

Note: *N* stands for number of observations. Weights have been used in all calculations.

Table 6.14 Percentage of firms paying in advance, 1993–95

	1993		*1994*		*1995*	
	%	*N*	%	*N*	%	*N*
Micro	0.00	44	7.30	41	2.90	43
Small	10.10	35	13.80	44	17.50	41
Medium	0.50	48	16.50	40	3.60	44
Large	15.80	35	21.20	40	21.60	34
Macro	0.00	5	0.00	4	8.90	4
Formal	4.90	122	15.80	123	10.60	120
Informal	3.30	45	9.30	45	6.80	46
African	3.50	71	9.80	72	7.10	72
Asian	0.50	83	8.50	84	5.80	81
Other	2.30	13	3.30	13	2.30	13
All	3.40	167	9.50	169	6.90	166

Note: *N* stands for number of observations. Weights have been used in all calculations.

A decline in the share of these firms providing trade credits is quite clear in the table. There is no similar decline for Asian firms and for relatively large firms.

Table 6.14, regarding firms paying in advance, does not change the picture. On the contrary, fewer African firms in the smallest category in particular must pay in advance. Furthermore, there is a substantial increase in advance payments by large and macro-firms.

Although Table 6.2b shows that trade credits and advance payments did not contribute a very large share in the total financing of firms, these forms of financing can play an important role for firms, since they provide working capital. The 1995 survey reveals that the working capital situation for small African firms in particular improved substantially.

6.6 Summary and policy implications

We drew the following policy conclusions in Isaksson and Wihlborg (1994, 1995) from studies of the financing of manufacturing firms in Kenya in 1993 and 1994:

1 The policy implication of these observations is that property rights to land for individuals need to be both expanded and strengthened in the sense that the owners should be able to transfer real property without consent from so-called Land Control Boards. These boards can veto the transfer of land to a bank after a borrower's failure to repay. The bankruptcy process requires improvement as well. The duration of the process and lack of faith in the legal system implies that lenders face great uncertainty about recovering loans.
2 Informal financial institutions do not seem to play a large role for manufacturing firms. It is possible that providing legal protection for informal financial transactions, in the form of court enforcement of contracts, could help the informal sector expand.
3 The importance of institutional reform, the strengthening of property rights legislation, and other aspects of the legal system, is heightened by the finding that firms with insufficient ability to pledge collateral for formal loans are further financially constrained by lack of access to trade credits, by having to offer trade credits themselves, and by needing to pay in advance.

Most of these policy conclusions still stand after including survey data for 1995 in the analysis. The limited access to, and the limited use of, formal loans seem to be explained by the need for firms to pledge collateral of a magnitude two or three times the loan. Most small firms simply do not have sufficient assets to access the loan market, and large firms are constrained in the amount of debt financing they can obtain. Only institutional development of the kind mentioned above may resolve these problems.

The weak legal structure in Kenya implies that implicit contractual relations and self-enforcing mechanisms of contracts, such as staking a firm's reputation, become important. Trade credits and advance payments as sources of financing, as well as access to informal loan markets, depend to a great extent on information about the borrower and self-enforcement of contracts. Information asymmetries between borrowers and lenders, in combination with an inability to enforce contracts, would lead to the exclusion of firms from all kinds of financing.

The positive aspects of the survey data for 1995 refer to the improved access to working capital financing in 1994 relative to the previous years. In particular very small and small African-owned firms seemed to have gained greater access to both trade credits and advance payments in 1994. At the same time the very large firms seemed to have increased their willingness to pay in advance. There remained a difference in access to these sorts of financing between Asian-owned and African-owned firms. Asian firms had greater access to credits, but the difference relative to African firms decreased in 1995.

Informal loans from moneylenders and other informal institutions remained unimportant in the manufacturing sector. Only angels among informal sources contributed significantly to the financing of firms. About 20 per cent of informal, African-owned, very small and small firms relied on angels.

Notes

1. When this chapter was written, Anders Isaksson was a research associate at the Department of Economics, Göteborg University, Sweden. He is now at UNIDO. He can be reached at aisaksson@unido.org. Note that the views expressed herein are those of the authors and not necessarily those of UNIDO.
2. This overview of the macroeconomic conditions between 1992 and 1995 to a large extent ignores a discussion of the several Enhanced Structural Adjustment Facility programmes implemented in Kenya. We feel that their influence is evident in the macro-data. How these programmes affect macro outcomes is, however, beyond the scope of this chapter.
3. Building societies are not included.
4. See the World Bank (1992, 1993).
5. The debt-equity ratio is very low by the standards of developing countries with functioning financial markets (Hussain, 1995).
6. Total assets include inventories, the replacement value of land, building, and equipment, informal lending, trade credits given, and advance payments given.
7. As noted, these few firms could be subsidiaries of multinationals.

References

Biggs, T., Raturi, M. and Srivastava, P. (1996), 'Enforcement of Contracts in an African Credit Market: Working Capital Financing in Kenyan Manufacturing', *RPED Discussion Paper*, World Bank, Washington, DC.

Eliasson, G., Rybczynski, T. and Wihlborg, C. (1993), *The Necessary Institutional Framework to Transform Formerly Planned Economies*, Industrins Utrednings Institut, Stockholm.

Hussain, Q. (1995), *'Banking and Equity Markets in Middle Income Countries'*, PhD dissertation, University of California, Berkeley, California.

International Monetary Fund (1998), *World Capital Markets*, IMF, Washington, DC.

Isaksson, A. and Wihlborg, C. (1994), 'The Financial Sector and Investments', Ch. 9 in *Limitations and Rewards in Kenya's Manufacturing Sector: A Study of Enterprise Development*, RPED, prepared for the World Bank, Africa Technical Department, by the Departments of Economics, Göteborg University and Nairobi University.

Isaksson, A. and Wihlborg, C. (1995), 'Financing and Investment', Ch. 5 in *Manufacturing in Kenya under Adjustment*, RPED, prepared for the World Bank, Africa Technical Department, by the Departments of Economics, Göteborg University and Nairobi University.

Isaksson, A. and Kulundu, D. (1998), 'Kenya 1997: From Chaos to Prosperity?', *Country Economic Report, 1998:4*, SIDA, Stockholm.

World Bank (1992), 'Tapping Kenya's Potential: A Private Sector Development Strategy', Public and Private Enterprise Division, Africa Region, World Bank, Washington, DC.

World Bank (1993), 'Performance Audit Report, Kenya: Financial Sector Adjustment Credit', World Bank, Washington, DC.

Part III

Investment Behaviour

7
Investment Behaviour

Måns Söderbom

7.1 Introduction

It has long been argued that fixed capital formation, or investment, in less developed economies is constrained by financial market segmentation, resulting in inefficient financial intermediation and misallocation of resources (McKinnon, 1973). This has been supported at the macro level by the empirical results reported by King and Levine (1993) and Easterly and Levine (1995), indicating that African growth has been hampered by poorly developed financial markets. Having experienced a series of financial crises in the late 1980s, Kenya launched a significant financial reform programme in 1989, aiming to enhance prudential controls, to ensure a more efficient allocation of credit, and, ultimately, to spur investment. Although some important steps, such as the liberalisation of interest rates, were taken in the early 1990s the poor investment record of the economy persisted (see Chapter 2 in this volume for a discussion of the sluggish investment record during 1990–95).

Why did investment not respond to these reforms? One possibility is that the reforms were not far-reaching enough, so that firms who wanted to turn to the financial market for funding of investment remained unable to do so. In this case firms would still be credit constrained, which, from a policy perspective, would warrant the continuation of extensive reforms of the credit market. An alternative explanation, however, could be that investment is mainly determined by non-financial factors, such as demand conditions or the cost of capital, allowing only a modest role for external finance. In this case, the firm would not be credit constrained, and policy steps targeted to increase investment should therefore be directed somewhere other than towards the financial system.

In this chapter we shall examine how financial factors have influenced investment in plant and machinery in Kenyan manufacturing over the 1992–94 period. Following the modelling strategy of Whited (1992) we derive an investment equation from the first-order conditions, where financial constraints are reflected in the Lagrange multipliers. In this framework, outlined in section 7.2, an increase in the internal liquidity for a constrained firm will effectively decrease its discount rate and hence increase investment. In section 7.3 we show descriptive statistics on investment and the role of finance. In section 7.4 we discuss the empirical formulation of the investment equation, and show regression results. We provide a summary and conclusions in section 7.5.

7.2 The theoretical model

In this section we seek to derive a theoretical investment model allowing for financial constraints. We make the standard assumption that the objective of the firm's manager(s) is to maximise the value of the firm, defined as the present discounted value of the expected future dividend streams:

$$V_t(K_{t-1}) = \max_{L(t),I(t)} E_t \sum_{s=t}^{\infty} \beta_t^{s-t} \pi_s \tag{7.1}$$

where subscripts on variables denote time, and K = capital, E = the expectations operator, $\beta = 1/(1+r)$, r = the discount rate, $\pi_s = P_s A_s F(K_s, L_s) - w_s L_s - G(I_s, S, P^I_s, Z) + B_s - (1 + i_{s-1})B_{s-1}$, P = the output price, A = total factor productivity, F = the production function, L = labour, w = the wage, G = the investment cost function, representing purchase and adjustment costs, I = investment, S = firm size, assumed time invariant, p^I = the purchase price of capital, Z = fixed adjustment cost, B = the value of outstanding debt, and i = the lending rate.

If there are no limits on borrowing and no transaction costs involved in issuing new equity, the firm's manager will choose investment and labour at time t in order to maximise (7.1) subject to the capital evolution constraint which we define as:

$$K_s = I_s + (1 - \delta)K_{s-1} \tag{7.2}$$

where δ is the depreciation rate of the capital stock. In this case the decision rule for optimal investment would be independent of financial variables. In general, however, assuming unlimited borrowing and costless transactions is not realistic, particularly not for sub-Saharan Africa where the financial infrastructure is poor.[1] In order to obtain a more relevant investment model allowing for financial imperfections, we thus need to adjust the theoretical framework. To this end we first drop the assumption of unlimited borrowing, and instead we assume that the firm has a debt ceiling:

$$B_s \leq B_s^* \tag{7.3}$$

This squares well with the Stiglitz and Weiss (1981) demonstration that creditors may voluntarily ration credit rather than increase the interest rate if there are principal-agent problems arising from asymmetric information (Vermeulen, 1998). Second, we assume that the firm cannot issue new equity, implying that dividends must be non-negative:

$$\pi_s \geq 0 \tag{7.4}$$

This will be realistic if current owners, for one reason or another, are unwilling to take additional owners on board, or if issuing new equity is prohibitively costly, perhaps due to transaction costs. Faced with these financial imperfections, the firm chooses investment and labour in order to maximise (7.1) subject to (7.2), (7.3) and (7.4).

To derive an investment equation suitable for empirical analysis, we need to make further assumptions. First, we assume that the production function $F(.)$ is Cobb-Douglas and homogeneous of degree one in capital and labour (i.e., constant returns to scale), which simplifies the analysis somewhat. The evidence reported in Chapter 8 in this volume suggests that this is not an overly strong restriction. Second, we assume that firms are price-takers in factor and goods markets, which also does not appear to be too restrictive given that most of the firms are quite small. Third, we write the investment cost function $G(.)$ in general form as

$$G(I_s, S, p_s^I, Z) = g(I_s, S, p_s^I) + 1_{\{I(S) \neq 0\}} \cdot S \cdot Z \tag{7.5}$$

where, for non-negative I, g is increasing and strictly convex in I, decreasing and strictly convex in the size variable S, and increasing in p^I. Further, Z is a fixed cost incurred whenever there is any investment and $1_{\{\xi\}}$ is an indicator function equal to one if the event ξ is true, and zero otherwise. The presence of fixed costs in (7.5) complicates the maximisation problem somewhat, since first-order conditions are not sufficient to guarantee optimum. This is because zero investment will be the best policy if the investment satisfying the first-order condition brings about fixed costs which are too high. However, when firms are price-takers with constant returns to scale (CRS) in production, which is assumed here, the first-order conditions guarantee optimal investment, provided that zero investment is not the optimal policy (see Abel and Eberly, 1994). Formally, this is represented by the following necessary and sufficient rule for positive investment:[2]

$$q_t I_t - G(I_t, S, p_t^I) > 0 \tag{7.6}$$

where q is the marginal valuation of a unit of installed capital. Given that it is optimal to invest, the optimal level will satisfy the first-order condition:

$$G_I'(t) = q_t$$

Hence, q affects both the decision to invest and the level of investment. In contrast, the fixed cost element in $G(.)$ only affects the decision to invest and not the investment level conditional on investment, since the marginal adjustment cost is independent of the fixed cost.

The next step is to unpack q in order to separate out its components. We show in Appendix 7.1 how, under our assumptions, the first-order condition (7.7) can be expressed as:

$$G_I'(I_t) = E_t \sum_{s=t}^{\infty} \left(\frac{1-\delta}{1+i_t}\right)^{s-t} \prod_{j=t}^{s} \left(1 - \frac{\gamma_j}{1+\lambda_j}\right) P_s A_s F_K'(K_s, L_s)$$

where γ_j is the Lagrangean multipliers on the debt constraint (7.3), λ_j is the multiplier on the dividend constraint (7.4), and where the lending rate is assumed constant for simplicity.[3] By definition, the right-hand side of (7.8) is equal to q. If the financial constraints (7.3) and (7.4) are expected never to bind, so that all expected γs and λs are zero, the associated product in (7.8) will be equal to one, yielding the standard neoclassical model in which investment is independent of finance. When the financial constraints bind, however (so that γ and λ are positive), investment is negatively affected since $E_t \Pi_j (1 - \gamma_j/(1+\lambda_j)) < 1$. Hence, because financial constraints increase the marginal opportunity cost of investment, a constrained firm behaves as if it has a higher discount rate, and thus a higher user cost of capital, than an otherwise identical firm which is financially unconstrained (Whited, 1992).

In order to translate (7.8) into an expression suitable for empirical analysis, we need an explicit form for the marginal product of capital. To this end we can utilise the implicit assumption that the firm in each point in time sets its employment level in accordance with the first-order condition with respect to labour, and write the marginal product of capital as:

$$F_K'(K_s, L_s) = b_1 w_s^{-b_2} (P_s A_s)^{b_2} \tag{7.9}$$

where b_1 and b_2 are composites of the input elasticity in the Cobb-Douglas production function. Further, we need to address the fact that expected future variables appearing in (7.8) are unobserved. We shall assume that these future variables can be written:

$$E_t X_s = X_t (1 + g_t^X)^{s-t} \tag{7.10}$$

where g_t^X is the expected growth rate of $X = \{P, A, w\}$. This follows Denny and Nickell (1992), with the exception that we use discrete rather than continuous time. For the integral in (7.8) to converge, these expected growth rates are required to be small relative to the discount factor. Third, for simplicity we assume that the firm holds static expectations about the state of financial constraints:

$$E_t \prod_{j=t}^{s} (1 - \gamma_j/(1+\lambda_j)) = (1 - \gamma_t/(1+\lambda_t))^{s-t} \equiv \Gamma_t^{s-t} \tag{7.11}$$

Using first-order expansions as described in Appendix 7.1, equation (7.8) can then be approximated as a log-linear form, yielding:[4]

$$\ln I_t = \alpha_0 + \alpha_1 \ln S + \alpha_2 \ln P_t + \alpha_3 \ln A_t - \alpha_4 \ln w_t + \alpha_5 g_t^{P} + \alpha_6 g_t^{A} - \alpha_7 g_t^{w} - \alpha_8 \Lambda_t - \alpha_9 i_t \quad (7.12)$$

where $\Lambda_t \equiv \gamma_t/(1 + \lambda_t)$ indicates the severity of financial constraints, and the α-coefficients are positive composites of the underlying structural parameters.[5] Hence, investment depends positively on output price, total factor productivity, their expected growth rates, and firm size, and negatively on the level and expected growth rate of wages, the interest rate and the degree of financial constraints. This equation will form the basis for the empirical analysis below. Before turning to the econometric analysis, however, we proceed in section 7.3 with a descriptive analysis of investment in general and the role of finance in particular.

7.3 Descriptive statistics

The empirical analysis in this chapter will be based on 450 observations associated with 216 firms. This sample, which we will refer to as the 'full sample', contains 85 firms which have been observed three times over the 1992–94 period, 64 firms which have been observed twice, and 67 firms for which there is only one observation. Observations with missing values or evidently incorrect data on key variables for the analysis have been eliminated according to principles discussed in Appendix 7.2, and we have also deleted any of the remaining firms which ever reported a replacement value of the capital stock as being less than Ksh20 000 in constant 1991 prices.[6] The reason for the latter rule is that the economic theory of investment behaviour proposed above is unlikely to provide a good approximation of the behaviour of very small firms. For more details about the sample selection, and variable definitions, see Appendix 7.2. General details on sampling procedures are provided in Chapter 3 in this volume. We will begin the empirical analysis by examining descriptive statistics of finance and investment. We will occasionally base our estimates on smaller sample sizes than that of the full sample, due to missing data. In Tables 7.1–4 sampling weights will be used.

A salient feature of the data is the large frequency of zero investments. In Table 7.1 we show the average proportions of firms making some investment in equipment or machinery during one year, split into size and sector groups. As can be seen in the far right column of the table, more than half of the firms in all size categories except the largest one refrain from investing during a whole year. Using a probit model with sample weights, where dummy variables for sector, size groups, and year enter as 'explanatory' variables, we can reject at the 5 per cent level of significance the hypothesis

Table 7.1 Average proportions of non-zero investments, by sector, status and size

	Food	*Wood*	*Textiles*	*Metal*	*All sectors*
Informal firms					
All informal	0.67	0.57	0.22	0.59	0.47
	7	*25*	*19*	*29*	*80*
Formal firms					
Micro ($L \leq 5$)	–	0.17	0.55	0.17	0.36
	0	*7*	*8*	*7*	*22*
Small ($5 < L \leq 20$)	0.29	0.38	0.54	0.45	0.39
	20	*18*	*21*	*18*	*77*
Medium ($20 < L \leq 75$)	0.27	0.54	0.52	0.52	0.41
	38	*42*	*32*	*33*	*145*
Large ($L > 75$)	0.76	0.66	0.44	0.67	0.67
	31	*33*	*29*	*33*	*126*
All formal	0.48	0.42	0.51	0.46	0.48
	89	*100*	*90*	*91*	*370*
F-test (1)			F(3, 192) = 3.84**		
F-test (2)			F(3, 192) = 0.33		
F-test (3)			F(3, 37) = 2.29*		

The F-tests are adjusted Wald tests, $F(k, d-k+1) \sim (d-k+1)W/kd$, where W is the Wald statistic, k is the number of restrictions, and d is the number of firms minus the total number of strata. To make these computations we used a probit with sampling weights to model the probability of investment, entering the relevant sector and size categories and wave dummies as explanatory variables. The test statistics are tests for the associated restrictions on these coefficients.

F-test (1) tests the hypothesis that for formal firms the proportion of non-zero investments is equal across the size groups.

F-test (2) tests the hypothesis that for formal firms the proportion non-zero investments is equal across sectors.

F-test (3) tests the hypothesis that for informal firms the proportion non-zero investments is equal across sectors.

Notes: Sampling weights were used in the calculations. Numbers in italics are number of observations. Significance at the 10% and 5% level is indicated by * and **, respectively. L denotes the number of employees.

that all four formal size groups have equal investment propensities (F-test 1).[7] This is driven by the higher propensity in the largest size group. Based on the same regression, we tested for sectoral differences and could not reject the hypothesis that investment propensities are equal across sectors (F-test 2). For informal firms there is evidence of sectoral differences, as indicated by the rejection of the null hypothesis at 10 per cent level of significance (F-test 3). This result is driven by the low degree of investment activity of textiles firms in the informal sector.

We proceed by investigating the distribution of net investment rates, defined as the investment minus capital sales, divided by the replacement value of capital. Figure 7.1 shows that investment activity is low, with approximately 75 per cent of the firms having investment rates smaller than 10 per cent. The distribution of investment rates is highly skewed,

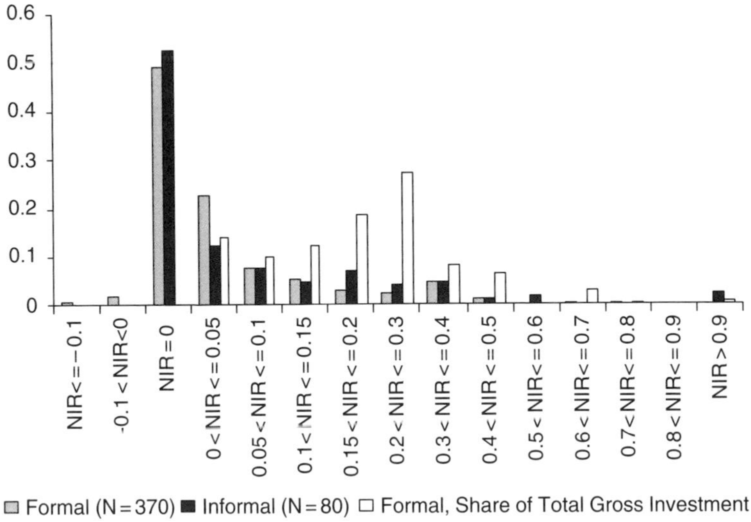

Figure 7.1 Frequency distribution of net investment rates and share of aggregate investment, formal and informal firms
Notes: *NIR* denotes the net investment rate. Sampling weights were used.

with very few firms selling off equipment as indicated by the low frequency of negative investment rates. Further, a large share of total investment expenditure is accounted for by a small fraction of the firms, as indicated by the 'share of aggregate' bar. Only 9 per cent of the formal firms have investment rates exceeding 20 per cent, but these investments constitute 45 per cent of total investment. This suggests that at the macro level, investment may to a large degree depend on the decisions of relatively few agents.

Next, we investigate irreversibility in a little more detail. The first column in Table 7.2 shows that it is not uncommon for these firms to purchase second-hand equipment, which may suggest that irreversibility is not an overwhelming problem after all. If firms can, and do, buy used capital stock, it also seems probable that they can sell off equipment if necessary. However, the propensity to purchase partly used equipment varies with firm size. Large firms, which typically manufacture more advanced products than small firms, are less likely to resort to the second-hand market for capital goods. One possible explanation for this result is that second-hand markets for advanced and/or idiosyncratic capital goods are shallower. The second column of Table 7.2 shows average ratios of the replacement to the sale value of the equipment. Clearly, there seem to be substantial wedges between the cost of purchasing and the price of selling, as mean ratios are around 0.5. Hence, solving, say, a temporary liquidity crisis by selling off

Table 7.2 Irreversibility indicators

	(1) Share of investments containing (partly) used equipment#	*(2) Mean ratio of sale to replacement value of equipment*	*(3) Share of firms selling equipment during one year*
Informal firms			
All informal	0.61	0.49	0.00
		(0.04)	
	25 [16]	*71 [9]*	*76 [4]*
Formal firms			
Micro	0.57	0.51	0.00
		(0.05)	
	5 [0]	*20 [2]*	*21 [1]*
Small	0.42	0.48	0.11
		(0.04)	
	23 [10]	*69 [8]*	*74 [3]*
Medium	0.36	0.43	0.03
		(0.02)	
	49 [16]	*131 [14]*	*133 [12]*
Large	0.18	0.53	0.07
		(0.04)	
	63 [18]	*118 [8]*	*118 [8]*
All formal	0.32	0.49	0.06
		(0.02)	
	140 [44]	*338 [32]*	*346 [24]*

Notes: # Based on subsample of investing firms.

Sampling weights were used in the calculations. Numbers in ( ) are standard errors, calculated without finite population correction, numbers in italics are number of observations, and numbers in [ ] are number of missing values.

See Table 7.1 for definition of firm size.

capital stock would typically imply that the firm would have to accept a fairly high discount. This presumably explains the very low share of firms ever selling equipment, as reported in the third column of Table 7.2. Of course, this may also make the firm reluctant to invest in the first place.

We now turn to aspects of investment funding. First, we examine the sources of investment financing, as reported in Table 7.3. Company retained earnings represent the primary source of finance, funding on average over 60 per cent of a firm's investment. For informal firms and formal firms with less than 75 employees, personal savings of the owner represent the second most important source (22 per cent and 10 per cent, respectively), yielding an overall reliance on internal funds of 86 per cent and 82 per cent, respectively, for these categories. The corresponding number for large formal firms is 66 per cent. As indicated by the F-test, the hypothesis that the averages for these three categories are equal can be rejected at the 10 per cent level, but not at the 5 per cent level. By

Table 7.3 Funding of investments

	Informal firms	*Formal firms*		
		L≤75	*L>75*	*All*
1 Company retained earnings	0.641	0.724	0.0664	0.698
	(0.078)	(0.073)	(0.069)	(0.050)
2 Personal savings	0.219	0.100	0	0.056
	(0.070)	(0.074)	–	(0.044)
3 Loan from friends/relatives	0.000	0.001	0.001	0.001
	–	(0.001)	(0.001)	(0.001)
4 Loan from bank	0.058	0.088	0.148	0.114
	(0.041)	(0.032)	(0.047)	(0.027)
5 Supplier credit	0.046	0.011	0.046	0.027
	(0.028)	(0.007)	(0.037)	(0.016)
6 Loan from moneylender	0.036	0.000	0.000	0.000
	(0.035)	–	–	–
7 Other	0.000	0.076	0.140	0.104
	–	(0.033)	(0.056)	(0.031)
Number of observations	34	94	77	171
Number of missing values	(7)	(9)	(4)	(13)
F-test		$F(2, 117) = 2.72^*$		

*Significance at the 10 per cent level.
Note: Sampling weights were used. The numbers in parenthesis are standard errors, calculated without finite population corrections. *L* denotes employment.

international standards these averages are atypically high, which can be explained by the fact that the economy has a shallow financial market.[8] This high degree of self-financing indeed suggests that investment could be sensitive to changes in liquid assets, as discussed in the theoretical section above. We examine this further in section 7.4.

The fact that firms to such a small extent use external funds could be because access is restricted or because the cost, relative to that of internal capital, is high. Restricted access can prevail in a liberalised system due to imperfect information, and would imply that firms underinvest because they are unable to realise their desired financial structure.[9] In this case, investment would be spurred by improved credit access. In contrast, if credit is available but at high cost, so that firms are not credit rationed, then investment would be unaffected by improved access. To shed some light on whether or not firms are rationed, we examine the data on application for formal loans. The results are reported in Table 7.4. We find that, during the course of one year, most firms do not apply for loans at all. Only 18 per cent of all formal firms apply, and 27 per cent of the informal firms.

Proceeding by examining the main reason for not applying, we find that a large share (70 per cent) of the non-applying formal firms state that loans are not needed. We take the responses 4–7 to indicate that the firm is close

Table 7.4 Loan applications

	Informal firms	*Formal firms*		
		$L \leq 75$	$L > 75$	*All*
Share of firms applying for loan in one year	0.27	0.15	0.25	0.18
Number of observations	76	237	128	362
Number of missing values	4	7	1	8
Main reason for not applying				
1 Process too difficult	0.15	0.03	0.00	0.02
2 Excessive collateral	0.25	0.02	0.02	0.02
3 Expected disapproval of application	0.12	0.02	0.00	0.01
4 Loan not necessary	0.15	0.71	0.67	0.70
5 Excessive interest rate	0.07	0.08	0.09	0.08
6 Reluctance to incur debt	0.12	0.03	0.11	0.05
7 Already heavily indebted	0.07	0.04	0.03	0.04
8 Other	0.07	0.08	0.10	0.08
Number of observations	54	192	87	279
Number of missing values*	2	4	5	9

* Referring to the sub-set of the sample not applying for a loan.
Note: Sampling weights were used.

to, or at, its desired financial structure, which is the case for 87 per cent of the non-applying formal firms. Reasons supportive of credit rationing, such as difficulties in the application process, and perhaps high collateral (though we note that high collateral is justified if the firm's capital stock is difficult to sell off in the case of a default on the loans) are thus perceived as the main reasons for very few firms. These results cast doubts on the idea that poor access is a binding problem. For informal firms the situation is somewhat different, since here 25 per cent of the firms state excessive collateral as the main reason for not applying for a formal loan. Furthermore, 15 per cent state that the application process is too difficult. Thus, credit access seems to be more of a problem for informal firms than for formal firms. Nevertheless, 41 per cent of the non-applying informal firms state one of 4–7 as the main reason, indicating that these firms have realised, or are close to reaching, their desired financial structure.

Summarising, in this section we have reported as a fundamental result that investment activity has been low in the majority of firms over the survey period. With this background, we have first shed light on the degree to which investment expenditures are 'sunk'. In particular, we take the low average ratio of the sale to the replacement value of equipment, and the very low share of firms selling off machinery, as indicators of sunk costs. Second, we have examined how firms finance their investments, and found a predominant role for internal funds. Since most firms who do not apply for formal loans appear to be close to their desired financial structure, the substantial reliance on internal funds signals high costs of external capital rather than credit rationing.

7.4 Empirical formulation and econometric results

The rest of this chapter will be devoted to econometric analysis. We will begin by discussing the specification of the empirical model, which will be based on the theoretical investment equation (7.12).

7.4.1 Model specification

First, to proxy the financial constraint term, Λ_t, in (7.12) we will focus on the liquidity position of the firm. Following Whited (1992), we showed in section 7.2 that for a firm operating under financial constraints, an increase in the internal liquidity will lower the effective discount rate because the credit constraints become less binding. Whited specified the financial-constraint element in the firm's discount rate as a function of the debt–equity ratio and the ratio of the firm's interest costs to the sum of interest costs and cash flow. The idea behind this is that the debt–equity ratio is a proxy for the firm's demand for borrowing, and the cost coverage ratio represents the likelihood of liquidity problems relative to the fundamental viability of the firm. Due to insufficient data on equity in the current data set, we are unable to calculate debt–equity ratios here. However, even with such data at hand there is some evidence (Bigsten *et al.*, 2000a) that it would be questionable to treat observed debt equity as an accurate measure of borrowing demand for the current sample.[10] Therefore we let the discount rate depend only on the cost coverage ratio, *COV*, along with industry control dummies:

$$\Lambda_t = c_0 + c_1 COV_t + \text{industry controls} \tag{7.13}$$

Since we focus on changes in internal funds as determining Λ_t it is important to incorporate all main sources of liquidity outflow. Hence we define the coverage ratio as the sum of wages, raw material costs and indirect costs, divided by the sum of profit, wages, raw material costs and indirect costs.[11]

Second, it is essential for estimation purposes to obtain an accurate proxy for the total factor productivity, ln(*A*). Nickell and Nicolitsas (1999) calculate this based on the Solow residual; however, the method is dynamic, which would be costly in our case due to the short span of the panel. Instead, we approach this problem by estimating production function directly. Assuming that TFP is time-invariant, which is probably not too restrictive given the short time span of the panel, the production function implicit in the derivation of the investment equation can be written:

$$\ln Y_{it} = \ln A_i + \alpha \ln L_{it} + (1-\alpha)\ln K_{it} \tag{7.14}$$

where *i* is the company indicator. In this case, it would in principle be possible to estimate the investment equation (7.12) allowing for fixed effects, without using auxiliary information from the production function. However, given the short (and unbalanced) panel and given the fact that only half of the observations on investment are positive, such procedures would yield substantial efficiency losses.[12] Also, measurement error bias would be enhanced. For the production function, however, we have complete observations on the dependent variable which implies that the problems with fixed effects discussed for the investment equation will be much less serious. In view of this we will use a two-stage approach, first estimating the production function allowing for fixed effects, and second using the estimated fixed effects from the production function as a proxy for ln(*A*) in the investment equation. Table 7.5 reports fixed effects production function estimates. In Model 1 we impose no restrictions on the returns to scale, which results in a very low coefficient on capital, and decreasing returns to scale. This appears to be a universal result for the RPED data sets (see Chapter 8 in this volume; Fafchamps and Oostendorp, 1999; Bigsten *et al.*, 2000b), which is probably driven, at least partially, by measurement errors in the capital stock. If these coefficients are downward biased, then the fixed effects will absorb scale and large (small) firms will on average appear more (less) productive than they are. Imposing constant returns to scale gives the results reported in Model 2. Of course if true returns to scale are decreasing, then the estimated TFP of large firms will be downward biased, and vice versa for small firms, when constant returns are imposed.

Although the above method is attractive because of its simplicity, it also has some problems. One problem is that the fixed effects estimates are inconsistent due to the short time horizon, which means that ln(*A*) will be

Table 7.5 Fixed effects estimates of Cobb-Douglas production function

	Model 1: Unrestricted returns to scale	*Model 2: Constant returns to scale*
Employment (log)	0.41***	0.86
	(0.11)	(–)
Capital (log)	0.06	0.14***
	(0.05)	(0.05)
F-test for fixed effects	4.26***	4.00***
Hausman specification test	22.2***	3.0
[N, NT]	216, 450	216, 450

Notes: The dependent variable is the logarithm of output.
Standard errors in parenthesis. Significance at 10%, 5%, and 1% level of significance is indicated by *, ** and ***, respectively.
Time dummies are included in the regressions.
Mean values and (standard deviations) of the variables used in estimation are, respectively: ln(Y(it)): 15.8 (2.30); ln(L(it)): 3.45 (1.60); and ln(K(it)): 15.4 (2.57).

measured with error. This implies that the coefficient on ln(A) in the investment regression will be inconsistent, the standard errors will be unreliable, and inferences should be made with care, just as in any other case where right-hand side variables are not perfectly measured. If we are prepared to assume random effects instead, we can estimate the production function and the investment equation simultaneously by maximum likelihood, integrating out ln(A). This procedure avoids introducing measurement errors, and it should also be more efficient. We will return to this when discussing the investment regressions.

Third, it is apparent from (7.12) that expectations are instrumental in determining investments. However, like most authors, we do not have explicit data on expected future output prices, productivity and wages in our data set. One solution to this problem is to generate the unobserved terms from subsidiary regressions (e.g., Gilchrist and Himmelberg, 1998; Nickell and Nicolitsas, 1999). Unfortunately, such an approach requires a fairly generous lag structure to be meaningful, which would be infeasible for us given the short panel. Instead we will assume that heterogeneity in expectations can be captured by firm-specific random effects which are uncorrelated with the regressors in the model.[13]

Recognising that our assumptions about expectations may nevertheless lead to misleading results if we are mistaken in formulating the theory, we proceed in the second part of the econometric analysis with a more direct treatment of expectations. We will use data collected in the last survey round only, on firms' expected percentage change in the sales volume in the subsequent year. We believe this is a reasonable proxy for whether the firm is optimistic about the future, which is essentially what the growth terms in (7.12) represent.

Fourth, firm size plays an important role in mediating the convex adjustment costs, which translates into a positive effect on the investment level in (7.12). Here we will use the replacement value of the capital stock in the year prior to the first observation of each firm (hence constant over time) to proxy size. Various control variables, such as industry sector and firm status, are used to capture variation in output and prices, and the average wage will be used for w.

Fifth, clearly the dependent variable in (7.12) will not be observed if the firm decides not to invest. It was discussed above how the fixed adjustment cost affects the decision to invest, but, conditional on investment, does not affect the investment level. This should be taken into account when specifying the empirical model for the decision to invest. However, since the fixed cost is unobserved we need to proxy it with observable variables. We expect search and managerial decision costs to constitute a considerable share of the fixed costs, and we therefore focus on factors linked to infrastructure, network participation, and organisational structure of the firms as potential proxies. After some elaboration, we have chosen to use dummy variables for location (infrastructure), ethnicity of the manager (Asian managers are known to be more inclined to participate in network activities), ownership structure, and legal status of the firm (organisational characteristics). We will model the decision to invest by a probit.

If unobserved factors in the probit regression are correlated with the unobserved factors in the primary equation (i.e., the equation for the investment expenditure), the parameter estimates will contaminated by sample selection bias (Heckman, 1976, 1979). Here we will assume that potential selection mechanisms are generated by time-invariant factors, for instance due to heterogeneity in expectations as discussed above. To illustrate the approach, we decompose the error terms in the two equations into a transitory and a permanent component:

$$\varepsilon_{1it} = u_{1it} + \mu_{1i} \qquad \text{(primary equation)} \qquad (7.15a)$$

$$\varepsilon_{2it} = u_{2it} + \mu_{2i} \qquad \text{(selection equation)} \qquad (7.15b)$$

Here u_{1it} and u_{2it} are assumed normally distributed with mean zero and standard deviation equal to σ and 1, respectively; μ_{1i} and μ_{2i} are assumed to follow a bivariate distribution; and u_{1it} and u_{2it} are taken to be independent of μ_{1i} and μ_{2i}, as well as mutually independent.[14] We will adopt a non-parametric strategy where the random effects are assumed to follow a discrete multinomial distribution. This distribution, which is bivariate, has $M \times K$ points of support, where the probability of observing the pair (μ_1^m, μ_2^k) is denoted PI_{mk}.[15] Hence the support points and the probabilities, which determine the entire distribution of the heterogeneity, are parameters to be estimated. This approach was suggested by Heckman and Singer (1984), and has subsequently been used in various microeconometric analyses of, for instance, dynamic discrete choice (Moon and Stotsky, 1993; Blau and Gilleskie, 1999), duration data (Blau, 1994; Ham and LaLonde, 1996), and count data (Deb and Trivedi, 1997). One of the merits of this strategy is flexibility: Monte Carlo evidence indicates that this approach compares favourably to standard Maximum Likelihood Estimation (MLE) correctly assuming normality, and that it performs better than MLE incorrectly assuming normality (Mroz and Guilkey, 1995; Mroz, 1999). Further details, and the likelihood function, are provided in Appendix 7.3.

Taking all of the above into account, our empirical version of equation (7.12) is as follows:

$$\ln I_{it} = \omega_1 - \omega_2 COV_{it} + \omega_3 \ln A_i + \omega_4 \ln K_{i0} - \omega_5 \ln w_{it} + controls_{it} + u_{1it} + \mu_{1i} \qquad (7.16)$$

where I = investment in equipment and machinery, COV = (raw material costs + wages + indirect costs)/(raw materials costs + wages + indirect costs + cash flow), A = TFP, K_0 = replacement value of the capital in the year prior to the first observation, w = average wage, *controls* = a set of dummy variables for industry and firm status (see Appendix 7.2); and where the ω-parameters are (positive) coefficients to be estimated. To model the decision to invest, we use the same explanatory variables as in (7.16) plus our proxies for fixed costs, denoted *FIXC*. Hence, the latent model can be written:

$$I_{it}^* = \tau_1 - \tau_2 COV_{it} + \tau_3 \ln A_i + \tau_4 \ln K_{i0} - \tau_5 \ln w_{it} + \tau_6 FIXC_{it} + controls_{it} + u_{2it} + \mu_{2i} \qquad (7.17)$$

where we observe

$$d_{it} = \begin{Bmatrix} 1 & if\, I_{it}^* \geq 0 \\ 0 & otherwise \end{Bmatrix}$$

where $d_{it} = 1$ indicates investment and $d_{it} = 0$ indicates no investment. Assuming that the residual, u_{2it}, follows a standard normal distribution, we can model the decision to invest using a probit.

We end this sub-section by showing in Table 7.6 sample statistics for the variables used in the regression analysis. Exactly half of the observations are zero investments. Our two measures of TFP, ln(A), are quite different. Estimating the production function without constraints gives higher values, and higher variation, than when we impose constant returns. It is possible that this is due to the freely estimated fixed effects absorbing firm scale, as discussed above. The average cost coverage ratio being 0.77 implies that on average 23 per cent of the value produced is retained in the firm as profits. This suggests that these firms are quite profitable.

7.4.2 Econometric results

In Table 7.7 we show selected parameter estimates for the investment equation (7.16).[16] In interpreting these results we shall mainly focus on the effects on investment conditional on investment being positive. That is, we will not refer much to the probit estimates, which are all reported in Appendix 7.4. Starting with the simplest model, the first column shows OLS estimates on the sub-sample of positive investments, using the freely estimated fixed effects from the production function as measures of TFP. Although the OLS procedure is restrictive in that it neglects the panel nature of the data as well as potential selectivity problems, it is useful as a benchmark.

The coefficient on total factor productivity, ln(A), is positive and highly significant. This coefficient is interpretable as the elasticity of investment with respect to total factor productivity, hence the point estimate of 0.57 suggests that investment might be quite responsive to variation in TFP.

Table 7.6 Sample statistics of variables in regression analysis

	All firms		*Investing firms*	
	Mean	*Standard deviation*	*Mean*	*Standard deviation*
Investment dummy*	0.50	0.50	–	
Investment (log)	–		12.69	2.53
ln *A*, free fixed effects	13.52	1.58	13.69	1.62
ln *A*, CRS fixed effects	10.67	1.02	10.70	0.98
Wage (log)	0.82	0.93	0.85	0.97
Cost coverage ratio	0.77	0.25	0.76	0.27
Food sector*	0.21	0.41	0.20	0.40
Wood sector*	0.28	0.45	0.29	0.46
Textiles sector*	0.24	0.43	0.21	0.41
Formal sector*	0.82	0.38	0.82	0.39
Small*	0.24	0.43	0.24	0.43
Medium*	0.33	0.47	0.29	0.46
Large*	0.28	0.45	0.36	0.48
Location in Nairobi*	0.65	0.48		
Asian manager*	0.58	0.49		
Legal status*†	0.41	0.49		
Private ownership*	0.18	0.38		
Expected sales change			0.19	0.26
NT, for all variables except sales expectation	450		225	
NT, for sales expectation			67	

* Dummy variables.
† Legal status is a dummy variable, equal to one if the legal status of the firm is sole proprietorship or partnership.

The estimated coefficient on the cost coverage ratio is equal to −0.60, and significant at the 10 per cent level. This coefficient is the negative of the semi-elasticity of investment with respect to the profit to output rate (see note 11), hence its magnitude means that a 1 per cent increase in the profit to output rate (i.e., a 1 per cent decrease in the cost coverage ratio) will be associated with a 0.60 per cent higher investment level, conditional on non-zero investment. This implies a rather low correlation between liquidity and investment: a firm having a fairly high profit rate of 0.20 is expected to invest only 6 per cent more than a firm with a small profit rate of 0.10, everything else being equal. The estimated coefficient on initial capital stock is 0.50 and significant at the 1 per cent level. This is not surprising given that the capital stock enters the empirical specification primarily to proxy firm size, whose presence in the theoretical investment equation derives from the adjustment cost function. The estimated wage coefficient is negative, as predicted by theory, but far from significant. In the probit, the coefficients of interest are

Table 7.7 Selected parameter estimates from four investment regressions

	1 *OLS and Probit*	2 *Panel selection model*	3 *Panel selection model*	4 *Panel selection model*
ln *A*, free fixed effects	0.57	0.52***		
	(0.12)***	(0.14)		
ln *A*, CRS fixed effects			0.34***	
			(0.13)	
ln *A*, one-step MLE				0.10
				(0.29)
Cost coverage ratio	−0.60*	−0.52	−0.66**	−0.72**
	(0.36)	(0.33)	(0.28)	(0.30)
Wage (log)	−0.03	0.13	0.17*	0.20**
	(0.11)	(0.11)	(0.10)	(0.10)
Initial capital (log)	0.50***	0.56***	0.73***	0.62***
	(0.06)	(0.06)	(0.04)	(0.05)
Food sector	1.00***	0.68***	0.61**	0.71***
	(0.26)	(0.23)	(0.24)	(0.24)
Wood sector	0.31	0.43*	0.46*	0.04
	(0.24)	(0.26)	(0.24)	(0.23)
Textiles sector	0.12	0.31	0.18	−0.16
	(0.26)	(0.26)	(0.24)	(0.22)
Formal firm	−0.41	−0.77**	−0.63**	0.33
	(0.31)	(0.36)	(0.30)	(0.37)
Wave 2	0.12	0.19	0.20	0.13
	(0.21)	(0.15)	(0.15)	(0.16)
Wave 3	0.09	0.09	0.08	0.09
	(0.21)	(0.16)	(0.16)	(0.16)
Intercept	−2.90***	−3.32***	−2.66**	2.25***
	(1.03)	(1.09)	(1.25)	(0.73)
Correlation (μ_1, μ_2)	–	0.12	0.13	−0.01
N, NT	216, 450	216, 450	216, 450	216, 450
Log Likelihood	−671.2	−622.4	−630.5	−1224.9

Notes: The dependent variable is the logarithm of investment. The omitted dummy category is an informal metal firm in wave 1.

Asymptotic standard errors in parenthesis. Significance at 10%, 5%, and 1% level of significance is indicated by *, ** and ***, respectively.

not very well determined, apart from that on total factor productivity which is positive and significant at the 5 per cent level.

How robust are these findings to controls for unobserved heterogeneity and sample selection? In Model 2 we report estimates from a panel selection model of the kind discussed in section 7.4.1, in which the random effects are being integrated out of the likelihood using a discrete mass points approach.[17] We retain the freely estimated fixed effect as our proxy for TFP. Compared to the estimates reported in Model 1, the coefficients on factor productivity and cost coverage are slightly smaller (in absolute terms). With a Z-statistic of 1.58, the cost coverage coefficient just fails to be significant at the 10 per cent level. The capital coefficient is slightly higher than before, and the wage coefficient now has the 'wrong' sign (i.e., positive) although it is still insignificant.

In the probit, factor productivity is even more significant than previously (now at the 1 per cent level), and the wage coefficient is negative and significant which is consistent with our theory. Our proxy variables for fixed costs – location in Nairobi, managerial ethnicity, firm status, and ownership – do not appear to work very well in the probit, as only the dummy variable for private ownership is (just) significant at the 10 per cent level. This suggests that the non-linearity of the model is important for identification.[18] It is clear that allowing for unobserved heterogeneity provides a significant improvement in the fit over Model 1, as the addition of the eight heterogeneity parameters increases the log likelihood value from −671 to −622. The estimated correlation between the two random effects is 0.12 which is not very high, suggesting that the two equations can be estimated independently (but with controls for unobserved heterogeneity) without giving results which are too misleading.

Model 3 is the same specification as Model 2, except that the fixed effects from the production function imposing CRS are used, instead of those from the freely estimated model. Although this yields a lower coefficient on total factor productivity, 0.34 compared to 0.52 in Model 2, the coefficient is still significant at the 1 per cent level. The coefficient on the liquidity terms is slightly higher (in absolute terms) than before, −0.66 compared to −0.52, and is significant at the 5 per cent level. Thus it appears that when we use the CRS fixed effects, more importance is attributed to liquidity at the expense of productivity in explaining investment. However, the liquidity effect is still not large: a firm with a profit rate of 0.20 would invest a mere 6.6 per cent more than a firm with a profit rate of 0.10, conditional on positive investment. The wage coefficient, which now is significant at the 10 per cent level, still has the wrong sign. As noted in section 7.2, a positive wage effect is inconsistent with a framework where the wage, from the firm's perspective, is fixed. However, one possible explanation for a positive wage effect (yet with the other effects in the model unchanged), proposed by Denny and Nickell (1992), is that the firm cannot alter its labour input in the short run unless it also alters the wage (e.g., by paying overtime premiums to workers). Alternatively, it may be that the wage variable picks up differences in firm efficiency not captured by our TFP measure. The initial capital stock now has a higher coefficient than previously, which suggests that the freely estimated fixed effects indeed absorb firm.

As discussed in section 7.4.1, the two-stage procedure on which we have relied so far, (i.e., first estimating the production function to obtain the fixed effects and then using these on the right-hand side of the investment regressions) is not free from problems. This is because the fixed effects are measured with error, due to the short span of the panel. An alternative approach, which avoids this particular problem, is to assume that the TFP terms are random, not fixed, effects, and then estimate the production function simultaneously with the investment equations. The idea is simply to write the likelihood for the three equations with ln(A) entering as a common factor, which means that we specify the residuals as:

$$\varepsilon_{1it} = u_{1it} + \mu_{1i} + \rho_1 \ln A_i \qquad \text{(primary equation)} \tag{7.18a}$$

$$\varepsilon_{2it} = u_{2it} + \mu_{2i} + \rho_2 \ln A_i \qquad \text{(selection equation)} \tag{7.18b}$$

$$\varepsilon_{3it} = u_{3it} + \ln A_i \qquad \text{(production function)} \tag{7.18c}$$

and then integrate over the distribution of the bivariate random effects (as in Models 2 and 3) and ln(A).[19] In this framework, the ρs in (7.18a) and (7.18b) have the same interpretation as ω_3 and τ_3 in (7.16) and (7.17) above (i.e., as coefficients on TFP in the investment equations). The transitory components of the error terms in (7.18a)–(7.18c) are assumed independent among themselves and of the permanent components. To facilitate estimation we shall also assume that ln(A) is independent of the μs. Estimation results are reported in Table 7.7, Model 4, and the most dramatic change concerns the productivity coefficient, as the estimated ρ_1 is 0.10 and insignificant.[20] The liquidity coefficient is slightly larger in magnitude than before (−0.72), but still implies a relatively modest liquidity effect on investment. Changes in the other coefficients are either small or unimportant for the discussion.

To summarise the results in Table 7.7, we have obtained point estimates on the cost coverage ratio ranging between −0.52 and −0.72, depending on the specification.[21] This indicates that investment is not very responsive to changes in liquidity. There are very few empirical studies based on African microdata to which we can compare this result; Bigsten *et al.* (1999), however, is one exception. The authors use data on Cameroon, Ghana, Kenya and Zimbabwe, and estimate an Accelerator model, regressing the investment rate (I/K) on the lagged profit to capital ratio and other (lagged) explanatory variables. They report a marginal effect of the lagged profit rate on the investment rate equal to 0.06.[22] To compare this to our estimated semi-elasticity, note that:

$$\frac{\partial(I/K)}{\partial(C/K)} = \frac{\partial(I/K)}{\partial \mathrm{I}} \frac{\partial I}{\partial(C/Y)} \cdot \frac{K}{Y}$$

In a steady state, when the firm invests in order to keep the capital stock constant over time, the first partial derivative on the right-hand side will be equal to ($1/K$). The second partial derivative is equal to the negative of our semi-elasticity multiplied by I. The median capital to output ratio in the sub-sample of investing firms is approximately 0.7. Hence, assuming $K/Y = 0.7$ and steady state, and setting the coefficient on our cost coverage ratio to −0.6, the marginal effect implied by our estimate is given by:

$$\frac{\partial(I/K)}{\partial(C/K)} = 0.6 \cdot 0.7 \cdot \frac{I}{K}$$

Hence, for investment rates less than about 0.14 (i.e., for most firms in this sample), the liquidity effect is lower than that reported by Bigsten *et al.*

We finally turn to the question of whether our assumption about expectations has implications for the results. As discussed in section 7.4.1, we have information on managers' expected percentage change in sales in the subsequent year. These data were only obtained in the last survey round and there are a number of missing values, which means that the sample size is small (67 firms) and the precision of the estimates is likely to be low. OLS results of the expectations-augmented investment equation are reported in Table 7.8. Because we would like to see how the estimate of the profit coefficient is affected by the presence of the expectation variable, we also report results from a regression without this variable. This is equivalent to assuming static expectations. Comparing columns 1 and 2, we see that the inclusion of the expectation variable has very minor effects on the parameter estimates. Thus there are no signs, from this small experiment, that the omission of an explicit expectations variable in the investment equation leads to seriously biased estimates.

Table 7.8 Selected OLS estimates for expectations-augmented investment equation

	1 *Static expectations*		2 *Self-reported expectations*	
	Coefficient	*Standard error*	*Coefficient*	*Standard error*
ln A, free fixed effects	0.78***	0.25	0.82***	0.26
Cost coverage ratio(t)	−0.69	0.70	−0.73	0.71
Wage(t) (log)	−0.20	0.26	−0.20	0.26
Initial capital (log)	0.42***	0.13	0.40***	0.14
$E(t)[\Delta\text{Sales}(t+1)]$			0.40	0.78
Adjusted R^2	0.70		0.70	
F-test	20.8***		18.3***	
[N, NT]	[67, 67]		[67, 67]	

Notes: The dependent variable is the logarithm of investment. Control variables for sector and firm status were included in the regression, but coefficients are not reported here to conserve space. Standard errors in parenthesis. Significance at 10%, 5%, and 1% level of significance is indicated by *, ** and ***, respectively.
The F-test tests H_0: all slope coefficients = 0.

7.5 Summary and conclusions

This chapter has analysed the investment behaviour of Kenyan manufacturing firms, with a focus on financial factors. Descriptive statistics indicate that investment has been low during the survey period, and that firms to a small extent rely on external finance when they invest. Judging from self-reported data it appears that most firms voluntarily refrain from formal borrowing. This suggests that it is factors other than restricted access to credit that explain low investment. We have also shown that sunk costs could be substantial, as indicated by the fact that few firms sell off equipment and by the low sale to replacement value of equipment. When firms faced by sunk costs choose to save for precautionary reasons rather than to invest, total investment will be hampered.

In the final part of the chapter we have estimated a neoclassical investment equation allowing for liquidity constraints. We have found a statistically significant role for liquidity, although the point estimates imply that investment is not overly sensitive to changes in liquidity: an increase in the profit to output ratio by 1 per cent is associated with an increase in investment by approximately 0.6 per cent. Whereas we have found that liquidity does not affect investment too much, our results do suggest that other factors, currently subsumed in the total factor productivity term, may be important in driving investment. One policy implication of these findings is that improving the availability of finance, typically a key element of financial reform programmes, is unlikely to have a major effect on investment. Policy efforts directed elsewhere, enabling firms to enhance total factor productivity for instance, may be more rewarding.

Appendix 7.1

Derivation of the theoretical investment equation

The firm's problem is to maximise (7.1), subject to (7.2), (7.3) and (7.4). Combining the first-order conditions with respect to I_t and K_t for this problem, we obtain the Euler equation:

$$G_I'(I_t) = P_t A_t F_K'(K_t, L_t) + E_t \rho_t \frac{1 + \lambda_{t+1}}{1 + \lambda_t} G_t'(I_{t+1}) \tag{A1.1}$$

where λ_s is the Lagrangean multiplier on the dividend constraint (7.4) (cf. equation (7) in Whited, 1992). This equation can be solved by forward substitution to yield:

$$G_I'(I_t) = E_t \sum_{s=t}^{\infty} \left\{ (\beta_t(1 - \delta))^{s-t} \prod_{j=t}^{s} \left(\frac{1 + \lambda_j}{1 + \lambda_{j-1}} \right) P_s A_s F_K'(K_s, L_s) \right\} \tag{A1.2}$$

where $(1+\lambda_t)/(1+\lambda_{t-1})=1$. The first-order condition with respect to debt B_t is

$$(1 + \lambda_t) - \gamma_t = (1 + i_t)\beta_t E_t(1 + \lambda_{t+1}) \tag{A1.3}$$

where γ_t is the Lagrangean multipliers on the debt constraint (7.4) (cf. equation (10) in Whited, 1992). Substituting (A1.3) into (A1.2) yields (7.8) in the text:

$$G_I'(I_t) = E_t \sum_{s=t}^{\infty} \left(\frac{1-\delta}{1+i_t}\right)^{s-t} \prod_{j=t}^{s} \left(1 - \frac{\gamma_j}{1+\lambda_j}\right) \mathrm{P_s A_s F_K'}(K_s, L_s) \tag{7.8}$$

Given the assumptions that technology is Cobb-Douglas, that expectations about future financial constraints are static, and that the expected values of $\{P, A, w\}$ are given by (7.10), we can rewrite (7.8) as:

$$G_I'(I_t) = E_t \sum_{s=t}^{\infty} \left(\frac{1-\delta}{1+i_t}\Gamma_t\right)^{s-t} ((1+g_t^P)^{1+b_4}(1+g_t^A)^{1+b_4}(1+g_t^w)^{-b_4})^{s-t} \times B_2 P_t^{1+b_4} A_t^{1+b_4} w_t^{-b_4} \tag{A1.4}$$

Provided that the growth terms in are small relative to the discount rate, and that agents hold point expectations, the integral will converge so that:

$$G_I'(I_t) = \frac{1+i_t}{1+i_t-(1-\delta)\Gamma_t(1+g_t^P)^{1+b_4}(1+g_t^A)^{1+b_4}(1+g_t^w)^{-b_4}} \times B_2 P_t^{1+b_4} A_t^{1+b_4} w_t^{-b_4} \tag{A1.5}$$

Hence:

$$\ln G_I'(I_t) = \ln(1+i_t) - \ln(1+i_t-(1-\delta)\Gamma_t(1+g_t^P)^{1+b_4}(1+g_t^A)^{1+b_4} \times (1+g_t^w)^{-b_4}) + \ln B_2 + (1+b_4)\ln P_t + (1+b_4)\ln A_t - b_4 \ln w_t \tag{A1.6}$$

The next step is to linearise the second term on the right hand side of (A1.6). For the growth terms we use the approximation $(1+x)^{\varphi} \approx 1+\varphi x$, hence:

$$\begin{aligned} &-\ln(1+i_t-(1-\delta)\Gamma_t(1+g_t^P)^{1+b_4}(1+g_t^A)^{1+b_4}(1+g_t^w)^{-b_4}) \\ &\approx -\ln(1+i_t-(1-\delta)\Gamma_t(1+(1+b_4)g_t^P)(1+(1+b_4)g_t^A)(1-b_4 g_t^w)) \\ &= -\ln(i_t+\delta+\Lambda_t-(1+b_4)g_t^P-(1+b_4)g_t^A+b_4 g_t^w + \text{cross terms}) \end{aligned} \tag{A1.7}$$

where $\Lambda_t = 1 - \Gamma_t$. Finally, we use the approximation $\ln(B+c) \approx \ln B + c/B$, which is accurate if c is small relative to B, and rewrite (A1.7) as

$$-\ln(\bar{i}+\delta) - \frac{\Lambda_t}{\bar{i}+\delta} + \frac{(1+b_4)}{\bar{i}+\delta}g_t^P + \frac{(1+b_4)}{\bar{i}+\delta}g_t^A - \frac{b_4}{\bar{i}+\delta}g_t^w - \theta i_t \quad \theta > 0$$

where we have linearised around the mean interest rate $\bar{i}$. Note that we have omitted the cross terms in (A1.7) since these are products of several small terms (and therefore of negligible magnitude). Using this equation in (A1.6) and solving for investment yields (7.12) in the text.

Appendix 7.2

The data

The original panel consists of 658 observations for 276 firms spanning the 1992–94 period. We delete observations with negative or missing values on value-added, and observations for which the sum of wages, costs of raw materials, and indirect costs exceeds the value of output. This leaves a sample of 527 observations. Furthermore, we delete observations with missing data on any of the following variables: investment in equipment, replacement value of equipment, employment, sector, location, ethnic origin of the owner, ownership structure, and legal status of the firm. This leaves a sample of 494 observations. Reported investments lower than Ksh2000 are set to zero.[23] Finally, we eliminate any of the remaining firms which ever reported a replacement value of the capital stock less than Ksh20 000 in constant 1991 prices.[24] The reason is that the economic theory of investment behaviour proposed in the text is unlikely to provide a good approximation for the decision rules of firms which are too small. This rule eliminates 45 observations on 23 firms from the sample, all of which are informal firms. The resulting sample, which we refer to as the 'Full Sample', has 450 observations on 216 firms.

Nominal values have been deflated using the appropriate deflators in Table A7.1, which were calculated from Central Bureau of Statistics (1995). See the appendix in Lundvall and Battese (2000) for details on these computations. We used output deflators for all variables except the capital stock and investment, where we use the capital deflator.

Table A7.1 Output and input deflators, 1991–94

	1991	*1992*	*1993*	*1994*
Output deflators by sector				
Food	100	114	154	188
Wood	100	112	149	176
Textiles	100	91	115	190
Metal	100	84	110	121
Bakery#	100	128	175	139
Capital deflator	100	105	129	174
CPI deflator	100	130	189	244

#Bakery is a sub-sector of the food sector. We follow Lundvall and Battese (2000) in using a separate deflator for the bakery sector, to account for the substantial fall in bakery prices in 1994.

Source: Calculated from Central Bureau of Statistics (1995).

The variables used in the empirical analysis are defined as follows.

1 Investment: Expenditure on investment in equipment.
2 Cost Coverage Ratio: calculated as sum of costs for raw materials, indirect costs and wages divided the sum of costs for raw materials, indirect costs, wages and cash flow.
3 Capital: Replacement value of equipment and machinery.
4 Wage: The wage bill, scaled by a factor of 10 000, divided by the number of employees.
5 Expected growth of sales: the expected percentage change in sales in the subsequent year. Data only available for wave 3.
6 Control Variables: industry sector dummies are equal to 1 if the firm belongs to the indicated sector, and zero otherwise. Solo or partnership, Formal, Asian manager, Location in Nairobi, and Private, are all dummy variables equal to one if the dummy category applies for the firm or observation, zero otherwise.

For the descriptive analysis in section 7.3 we use sampling weights calculated as the number of firms in each stratum in the population divided by the number of firms in each stratum in the sample. Criteria for stratification were sector (four categories), formal status (two categories), and, for formal firms, firm size (five categories). Hence, a total of 24 strata exist. When pooling the three waves we treat each firm as a cluster. For two reasons some of the results reported in Section 7.3 are based on fewer observations than the full sample of 450 observations. First, data may be missing at random, in which case we recalculate the weights in order to ensure that the population is still well represented. On the other hand, data may also be missing because the question is inapplicable for the firm (which will be the case for certain follow-up questions), in which case we do not recalculate the weights. The reason is that such information only exists for a sub-set of the population, and by keeping the sampling weights the sub-sample becomes representative of precisely this sub-set.

Appendix 7.3

The likelihood function for Models 2–4 in Table 7.7

Models 2–3

Conditional on the random effects μ_{1i} and μ_{2i} the likely contribution of firm i, observed during $t = 1,\ldots,T$, is equal to:

$$L_i(\mathbf{B}|\mu) = \prod_{t=1}^{T}\left\{\{\Phi(X_{2it}\beta_2 + \mu_{2i})\phi((y_{it} - X_{1it}\beta_1 - \mu_{1i})/\sigma_1)/\sigma_1\}^{d(it)} \times \{1 - \Phi(X_{2it}\beta_2 + \mu_{2i})\}^{1-d(it)}\right\} \quad \text{(A3.1)}$$

where **B** denotes the vector of parameters in the model, and X_{2it} = explanatory variables in the selection equation (probit); β_2= coefficients on X_{2it}; X_{1it} = explanatory variables in the primary equation; β_1 = coefficients on X_{1it}; y_{it} = dependent variable in the primary equation (log investment); σ_1 = standard deviation of transitory residuals in primary equation; $d(it)$ = dummy variable equal to one if there is investment and zero if not; $\Phi(\cdot)$ = cumulative distribution function of the standard normal distribution; $\phi(\cdot)$ = density of the standard normal distribution.

The unconditional likelihood is obtained by integrating over the bivariate discrete distribution of the random effects:

$$L_i(\mathbf{B},PI) = \sum_{k=1}^{K}\sum_{m=1}^{M} PI_{mk}\, L_i\big(\mathbf{B}\,|\,\mu_{1i} = \mu_1^m, \mu_{2i} = \mu_2^k\big) \tag{A3.2}$$

where PI is the vector of probabilities (PI_{11}, PI_{21},...,PI_{M1}, PI_{12},...,PI_{MK}) of the $M \times K$ support points of the heterogeneity distribution. The sample likelihood is the product of such terms over the sample. This formulation underlies Models 2 and 3.

Model 4

In Model 4 we amend the likelihood contribution of the production function to the expression in (A3.1), letting ln(A) be a common random factor in all three equations:

$$\begin{aligned} L_i(\mathbf{B}\,|\,\mu) = \prod_{t=1}^{T} \Big\{ & \{\Phi(X_{2it}\beta_2 + \mu_{2i})\phi((y_{it} - X_{1it}\beta_1 - \mu_{1i} - \rho_1 \ln A_i)/\sigma_1)/\sigma_1\}^{d(it)} \\ & \times \{1 - \Phi(X_{2it}\beta_2 + \mu_{2i} + \rho_2 \ln A_i)\}^{1-d(it)} \\ & \phi((Q_{it} - X_{3it}\beta_3 - \ln A_i)/\sigma_p)/\sigma_p \Big\} \end{aligned} \tag{A3.3}$$

where X_{3it} = explanatory variables in the production function; β_3 = coefficients on X_{2it}; Q_{it} = dependent variable in the primary equation (log output); ρ_j = factor loading, j = 1,2; and P-subscripts on the normal density and the standard deviation indicate that these refer to the production function. The unconditional likelihood is equal to

$$L_i(\mathbf{B},PI,\pi) = \sum_{r=1}^{R}\sum_{k=1}^{K}\sum_{m=1}^{M} \pi_r \times PI_{mk} L_i\big(\mathbf{B}\,|\,\mu_{1i} = \mu_1^m, \mu_{2i} = \mu_2^k, \mu_{3i} = \mu_3^r\big) \tag{A3.4}$$

where π is the vector of probabilities (π_1, π_2,...,π_R) of the R support points of the distribution of $\mu_3 \equiv \ln A$ and for notational convenience we let $\mu_{3i} \equiv \ln A_i$. The sample likelihood is the sample product of such terms. This formulation underlies Model 4.

Since we include intercepts in each equation only $U/\mu 1$ support points are identified, where U = R, K, M. Therefore we parameterise the support points as, $\mu_s^c = \theta_c W_{cs}$, where θ_c is a scale factor, c = 1,2,3 is the equation index, and

$$W_{cs} = \begin{cases} -0.5 & \text{if } s = 1 \\ 0.5 - (1 + \exp(-a_{cs}))^{-1} & \text{if } 1 < s < U \\ 0.5 & \text{if } s = U \end{cases} \quad \text{(A3.5)}$$

Further, the probability terms are required to be non-negative and to sum to unity, which is accomplished by specifying appropriate boundary and linear equality constraints available in the computer programme.

The semiparametric maximum likelihood estimates are found by maximising the sample log likelihood, $\log L = \sum_n \log L_n(\cdot)$, with respect to its arguments. To this end we used the SAS/IML Newton–Raphson Ridge Optimisation Method (NLPNRR). The estimation exercise is quite costly, since convergence is slow and may occur at a local optimum. To guard against convergence at local points, several different start values for each model were used.

Appendix 7.4

Table A7.2 Complete results for models in Table 7.7

	1 OLS & Probit		*2 Panel selection model*		*3 Panel selection model*		*4 Panel selection model*	
	Coef.	*Std err.*	*Coef.*	*Std err.*	*Coef.*	*Std err.*	*Coef.*	*Std err.*
A. Investment equation (dependent variable: log investment)								
Intercept	−2.90***	1.20	−3.32***	1.09	−2.66**	1.25	2.25***	0.73
ln A, free FE	0.57***	0.14	0.52***	0.14				
ln A, CRS FE					0.34***	0.13		
Wage (log)	−0.03	0.12	0.13	0.11	0.17*	0.10	0.20**	0.10
Cost coverage	−0.60*	0.33	−0.52	0.33	−0.66**	0.28	−0.72**	0.30
Food	1.00***	0.28	0.68***	0.23	0.61**	0.24	0.71***	0.24
Wood	0.31	0.23	0.43*	0.26	0.46*	0.24	0.04	0.23
Textiles	0.12	0.27	0.31	0.26	0.18	0.24	−0.16	0.22
Formal firm	−0.41	0.31	−0.77*	0.36	−0.63*	0.30	0.33	0.37
Initial capital (log)	0.50***	0.07	0.56***	0.06	0.73***	0.04	0.62***	0.05
Wave 2	0.12	0.20	0.19	0.15	0.20	0.15	0.13	0.16
Wave 3	0.09	0.22	0.09	0.16	0.08	0.16	0.09	0.16
B. Investment selection equation								
Intercept	−2.25**	0.89	−1.83	1.55	−0.35	1.73	1.38	0.90
ln A, free FE	0.17**	0.08	0.32***	0.12				
ln A, CRS FE					0.17	0.14		
Wage (log)	−0.05	0.08	−0.20*	0.12	−0.17	0.12	−0.18	0.12
Cost coverage	−0.11	0.26	−0.46	0.39	−0.60	0.37	−0.57	0.38
Food	−0.20	0.19	−0.45	0.35	−0.46	0.40	−0.46	0.33
Wood	0.06	0.17	0.25	0.25	0.19	0.25	0.03	0.25
Textiles	−0.22	0.18	−1.05**	0.53	−1.21***	0.43	−1.39***	0.47
Formal firm	−0.32	0.23	−1.18***	0.39	−1.18***	0.37	−1.13***	0.39

Table A7.2 continued

	1 OLS & Probit		*2 Panel selection model*		*3 Panel selection model*		*4 Panel selection model*	
	Coef.	*Std err.*	*Coef.*	*Std err.*	*Coef.*	*Std err.*	*Coef.*	*Std err.*
Initial capital (log)	0.00	0.04	−0.06	0.06	0.01	0.06	0.02	0.06
Location Nairobi	0.21	0.14	0.29	0.30	0.37	0.25	0.49**	0.24
Asian manager	0.06	0.15	0.27	0.27	0.35	0.31	0.36	0.26
Solo/partnership	0.20	0.18	0.02	0.36	−0.12	0.29	−0.22	0.28
Private firm	0.13	0.17	0.44*	0.27	0.47*	0.25	0.48*	0.26
Wave 2	0.29**	0.15	0.30	0.19	0.32*	0.19	0.32	0.19
Wave 3	0.03	0.14	−0.26	0.21	−0.26	0.20	−0.27	0.21
C. Production function (dependent variable: log output)								
Intercept							10.55***	0.52
Capital (log)							0.16***	0.04
Employment (log)							0.80***	0.06
Food							0.07	0.16
Wood							−0.78***	0.15
Textiles							−0.44***	0.16
Formal firm							0.22	0.18
Location Nairobi							0.34***	0.11
Asian manager							0.17	0.13
Solo/partnership							−0.65***	0.15
Private firm							0.16	0.14
Wave 2							−0.01	0.09
Wave 3							0.01	0.09
D. Auxiliary parameters								
$\sigma_{\text{Investment}}$	1.29	(—)	0.85***	0.07	0.87***	0.05	0.88***	0.06
$\sigma_{\text{Production}}$							0.78***	0.03
θ_1			3.11***	0.53	3.64***	0.34	3.99***	0.32
θ_2			3.07***	0.43	3.09***	0.43	3.19***	0.47
ln A, random eff.							1.07***	0.09
ρ_1							0.10	0.29
ρ_2							0.42	0.29
a_1			0.18	0.29	0.08	0.21	−0.01	0.18
PI11			0.04	0.14	0.01	0.05	0.02	0.03
PI21			0.54***	0.15	0.60***	0.08	0.61***	0.06
PI31			0.12*	0.06	0.07	0.04	0.05	0.03
PI12			0.04*	0.02	0.05**	0.02	0.06***	0.02
PI22			0.15***	0.04	0.16**	0.04	0.19***	0.04
PI32			0.12	–	0.11	–	0.07	–
π_1							0.51***	0.11
π_2							0.49	–
Log likelihood				−622.4		−630.5		1224.9

Coef. = coefficient.
Std err. = standard error.
Notes: Standard errors are asymptotic. Significance at 10%, 5%, and 1% level of significance is indicated by *, ** and ***, respectively.

Notes

1. See Ch 2 and 3 in Isaksson (1997) for a macroeconomic analysis of the financial structure in Kenya.
2. See Abel and Eberly (1994).
3. Equation (7.8) is obtained by solving the Euler equation (A1.1) in Appendix 7.1 by forward substitution.
4. This involves approximating the log of the expectation by the expectation of the log. Accuracy of this procedure requires variation to be small.
5. Due to lack of data, notably on prices, we will not be able to identify the structural parameters from our empirical estimates.
6. This corresponded to approximately US$700 in 1991.
7. Informal firms were excluded from the sample for this regression.
8. Assuming that the investing firm wants to keep the debt–equity structure constant and abstracting from tax considerations, the percentage by which an investment is funded by internal means can be interpreted as a flow equivalent to the equity to total assets ratio (i.e., solidity, in its simplest form). Based on the Global Vantage database, Rajan and Zingales (1995) report that average ratios of equity to total assets for non-financial firms range from 0.28 to 0.43 in the G-7 countries.
9. See the seminal paper by Stiglitz and Weiss (1981).
10. Bigsten *et al.* (2000a), in a cross-country study of credit constraints in African manufacturing, report that a non-trivial number of the firms with zero outstanding debt nevertheless have a demand for credit. The authors advance a number of reasons why credit demand is not met in these cases, e.g., due to high transaction costs in the credit application process, or credit rationing.
11. Note that this definition of *COV* is equal to one minus the cash flow to output ratio. We have experimented with a few variants of this formulation, excluding wage costs and/or costs of raw material. This produced only minor changes to the empirical results reported below.
12. Also note that firms with only one positive data point on investment will be excluded from estimation in first differences or differences from means.
13. This orthogonality assumption does not seem too restrictive, given that the random effects are thought to represent growth terms while the regressors are variables in levels.
14. It is straightforward to allow for a time variant selectivity effect in (7.15a)–(7.15b) as well. We neglect this here due to the short time span of the panel.
15. Hence $M = 1$, $K = 1$ gives the standard model without firm effects or selectivity.
16. Full set of results is shown in Appendix 7.4.
17. The results reported here are based on three mass points in the main equation and two in the probit. Increasing the number of mass points in any of the equations led only to minor improvements in the likelihood value.
18. To assess how sensitive the results are to such 'weak' identification, we experimented with using the lagged dependent dummy as an additional explanatory variable in the probit. This leads to additional complications because of the initial conditions problem (see, e.g., Heckman, 1981), but it has the advantage of providing additional identification. Indeed, according to our theory the lagged dependent dummy should not enter the model, but it is nevertheless highly significant when included. Most importantly, however, the results in the main equation are not affected much by the inclusion of the lagged dependent

dummy in the selection equation. Hence we would tentatively conclude that the results reported in Table 7.7 are not sensitive to the identification being quite weak. Because of this, and due to the additional methodological difficulties, we opted to proceed without using the lagged dependent dummy in the selection equation despite its significance.

19. In estimation we use the current value of the capital stock in the production function. Since current capital subsumes current investment, this is strictly not consistent with the econometric model as this relies on the system being recursive. However, this will not yield any major problems since the share of current capital contributed by current investment is going to be very minor due to the low investment activity. Nevertheless, we re-estimated Model 4 using the lagged instead of the current capital stock and the results were almost identical to those reported in Table 7.7. A related potential problem is that the capital stock is the outcome of past investment decisions, implying that there will be an endogeneity problem if past investments were affected by the same firm effects as current (during the span of the panel) investments. Remedying this appears to be quite complicated, however, and we made no such attempt.
20. For simplicity, we used only two support points in the production function.
21. To explore if firms of different sizes respond differently to changes in liquidity, we defined an interaction term as the product of the capital stock and the liquidity variable, and included this in the investment regressions. The associated coefficient was never significant, however, suggesting that there is little heterogeneity in the liquidity effect over firm size.
22. See Table 7, column (1), in their paper.
23. This corresponded to approximately US$70 in 1991.
24. Approximately equal to US$700 in 1991.

References

Abel, A. B. and Eberly, J. (1994), 'A Unified Model of Investment of Under Uncertainty', *American Economic Review* 84.

Bigsten, A., Collier, P., Dercon, S., Gauthier, B., Gunning, J. W., Isaksson, A., Oduro, A., Oostendorp, R., Pattillo, C., Söderbom, M., Sylvain, M., Teal, F., and Zeufack, A. (1999), 'Investment in Africa's Manufacturing Sector: A Four Country Panel Data Analysis', *Oxford Bulletin of Economics and Statistics* 61.

Bigsten, A., Collier, P., Dercon, S., Fafchamps, M., Gauthier, B., Gunning, J. W., Oduro, A., Oostendorp, R., Pattillo, C., Söderbom, M., Teal, F. and Zeufack, A. (2000a), 'Credit Constraints in African Manufacturing', mimeo, Oxford University.

Bigsten, A., Collier, P., Dercon, S., Fafchamps, M., Gauthier, B., Gunning, J. W., Oduro, A., Oostendorp, R., Pattillo, C., Söderbom, M., Teal, F. and Zeufack, A. (2000b), 'Risk Sharing in African Labour Markets', mimeo, Free University.

Blau, D. M. (1994). 'Labor Force Dynamics of Older Men', *Econometrica* 62.

Blau, D. M. and Gilleskie, D. B. (1999), 'Retiree Health Insurance: the Labor Force Behavior of Older Men in the 1990s', mimeo, University of North Carolina at Chapel Hill.

Central Bureau of Statistics (1995), *Statistical Abstract 1995*, Office of the Vice-President and Ministry of Planning and National Development, Nairobi.

Deb, P. and Trivedi P. K., (1997), 'Demand for Medical Care by the Elderly: A Finite Mixture Approach', *Journal of Applied Econometrics* 12.

Denny, K. and Nickell, S. J. (1992), 'Unions and Investment in British Industry', *The Economic Journal* 102.

Easterly, W. and Levine R., (1995), 'Africa's Growth Tragedy', mimeo, The World Bank, Washington, DC.

Fafchamps, M. and Oostendorp, R. (1999), 'Investment', ch. 6 in J. W. Gunning and R. Oostendorp (eds), 'Industrial Change in Africa: Micro Evidence on Zimbabwean Firms under Structural Adjustment', unpublished manuscript, Free University, Amstendam.

Gilchrist, S. and Himmelberg, C. P. (1998), 'Investment, Fundamentals and Finance', Working Paper 6652, National Bureau of Economic Research.

Ham, J. C. and LaLonde, R. J. (1996), 'The Effect of Sample Selection and Initial Conditions in Duration Models: Evidence from Experimental Data on Training', *Econometrica* 64.

Heckman, J. (1976) 'The Common Structure of Statistical Models of Trunctation, Sample Selection, and Limited Dependent Variables and a Simple Estimator for Such Models', *Annals of Economic and Social Measurement* 5.

Heckman, J. (1979), 'Sample Selection Bias as a Specification Error', *Econometrica* 47.

Heckman, J. (1981), 'The Incidental Parameters Problem and the Problem of Initial Conditions in Estimating a Discrete Time-Discrete Data Stochastic Process', in C.F. Manski and D. McFadden (eds), *Structural Analysis of Discrete Data with Econometric Applications*, MIT Press, Cambridge, MA.

Heckman, J. and Singer, B. (1984), 'A Method for Minimizing the Distributional Assumptions in Econometric Models for Duration Data', *Econometrica* 52.

Isaksson, A. (1997), 'Essays on Financial Liberalisation in Developing Countries: Capital Mobility, Price Stability and Savings', PhD Thesis, Göteborg University.

King, R. G. and Levine, R. (1993), 'Finance, Entrepreneurship and Growth: Theory and Evidence', *Journal of Monetary Economics* 32.

Lundvall, K. and Battese, G. E. (2000), 'Firm Size, Age and Efficiency: Evidence from Kenyan Manufacturing Firms', *Journal of Development Studies* 36.

McKinnon, R. I. (1973), *Money and Capital in Economic Development*, Brookings Institution Washington, DC.

Moon, C. and Stotsky, J. G. (1993), 'The Effect of Rent Control on Housing Quality Change: A Longitudinal Analysis', *Journal of Political Economy* 101.

Mroz, T. (1999), 'Discrete Factor Approximations in Simultaneous Equation Models: Estimating the Impact of a Dummy Endogenous Variable on a Continuous Outcome', *Journal of Econometrics* 92.

Mroz, T. and Guilky, D. (1995), 'Discrete Factor Approximations For Use in Simultaneous Equation Models with Both Continuous and Discrete Endogenous Variables', mimeo, University of North Carolina at Chapel Hill.

Nickell, S. and Nicolitsas, D. (1999), 'Does Innovation Encourage Investment in Fixed Capital?', mimeo, Centre for Economic Performance at London School of Economics.

Rajan, R. G. and Zingales, L. (1995), 'What Do We Know about Capital Structure? Some Evidence from International Data', *Journal of Finance* 50.

Stiglitz, L. and Weiss, A. (1981), 'Credit Rationing in Markets with Imperfect Information', *American Economic Review* 71.

Vermeulen, P. (1998), 'Detecting the Influence of Financing Constraints on Fixed Investment', mimeo, Stanford University, California.

Whited, T. M. (1992), 'Debt, Liquidity Constraints, and Corporate Investment: Evidence from Panel Data', *Journal of Finance* 47.

Part IV

Performance of Manufacturing

8
Productivity and Technical Efficiency

Karl Lundvall, Walter Ochoro and Lennart Hjalmarsson

8.1 Introduction

This chapter analyses the performance of the four manufacturing sectors included in the survey in terms of technical efficiency and productivity. The long-term trends of these variables are analysed in Chapter 2: here we are primarily interested in their behaviour during the survey years, and how they vary across sectors and among different categories of firms. We will also attempt to unveil some of the underlying factors behind the variation in performance. Since our analysis is confined to Kenyan firms alone, we cannot say anything conclusive about the impact of economy-wide factors such as corruption, infrastructure, institutional inefficiency and low demand simply because these are more likely to affect the overall *level* of performance, and not its variation among firms. Our focus is therefore on firm-specific variables, including location, age, growth, human capital, export behaviour, credit access and ownership. We base the selection of these factors on both theoretical and Kenya-specific empirical grounds. For policy considerations, the relative importance of these variables are of interest, and we draw some conclusions with this in mind in the final section of the chapter.

The next section contains a brief overview of the main efficiency constraints for manufacturing production in Kenya since Independence. Section 8.3 presents the variables used in the empirical analysis in the later sections and evaluates partial factor productivities for different firm sizes. Econometric estimations of technical efficiency using fixed effects and stochastic frontier models are conducted in section 8.4, which suggest considerable variation in technical efficiency and productivity across firms. Potential determinants for firm-specific productivity are discussed in section 8.5 and econometrically analysed in section 8.6 using average production functions. The chapter ends with a short summary and some conclusions in section 8.7.

8.2 Background

The rise in the share of manufacturing in total output and employment and a corresponding decline in agriculture is usually considered necessary for economic development of countries such as Kenya. The industrial structure that evolved in Kenya was, however, geared to the small domestic market with some limited targeting of the regional market in the Eastern African region. Import substitution was therefore the main focus of industrial policy from the colonial period through to the first two post-colonial decades and thus prescribed the scope of the industrial development that occurred.

Both the colonial and the post-colonial governments promoted the domestic manufacturing sector in a variety of ways. The structure of taxes and several price regulations, ranging from wages to input price policies, favoured the manufacturing sector. A number of financial and development organisations were also established to foster the growth of industry. The post-colonial government went even further, and chose to invest in manufacturing establishments. The governments of both regimes pursued protective trade policies to promote the establishment of domestic industries. Both tariffs and quantitative restrictions were applied. To attract foreign capital, a wide range of incentives was also instituted. Thus an inward-oriented industrial structure emerged. That such protection results in inefficient industries characterised by low productivity due to the absence of competition is an established fact (see Chapter 9 by Granér and Isaksson). The ability of the industries to penetrate export markets was consequently constrained.

The rate of expansion of the industry therefore became dependent on the growth of domestic demand. The limited size of the domestic market meant that scale economies, where these were technologically feasible, could not be realised. As the constraints to industrial growth due to the limits of the domestic market became clearer to policy-makers, the need for export orientation was appreciated and increasingly promoted. The domestic industries, which had evolved in such a protected environment, were ill-prepared to face the competition in the world markets. The trade liberalisation policies, followed under the World Bank- and IMF-sponsored structural adjustment programmes dating from the 1980s (which meant that the domestic manufacturers faced increasing competition from imports) made this abundantly clear. This is precisely the key problem facing the manufacturing sector. The low productivity that characterised most industries in the previous protective regimes left the sector ill-prepared for the new competitive pressures in the post-liberalisation period. Low productivity translates into high unit costs, which explain the competitive disadvantage in both the domestic and foreign markets.

Low productivity in the country has various underlying factors. The principal ones are sub-optimal plant sizes, outmoded technologies relative to those in use by competitors and underutilisation of installed capacities.

The appropriate post-liberalisation industrial strategy entails transformation from the historical inward to an outward orientation, with exports as the driving force. This is the core of Kenya's current industrial policy. It holds the promise of access to unlimited export demands for such a small country, and thus the removal of demand-side constraints to industrial growth, full capacity utilisation, technological advances, investment and growth. The export promotion hypothesis also has an efficiency component: exporting firms must be competitive, implying high technical efficiency and productivity relative to international standards. In a way, the RPED data enable us to test this hypothesis explicitly as sectors with varying degree of export content of output are surveyed. For instance, the food sector has traditionally been the major export sector in comparison to wood, textiles and metal. Hence, if food is more productive than these other sectors, the efficiency argument of exporting is supported.

Internally, the firms have to overcome the inefficiencies that developed in the protective regime to be able to expand into the larger export markets. The need for efficiency has become even more urgent also to maintain their traditional domestic markets against competing imports as the protective walls are removed in an increasingly liberalised imports regime. Efficiency is therefore important both for firms aiming to penetrate the export markets, and for those whose objectives are limited to the domestic markets. To survive, the firms will have to develop their competitive strengths to match their foreign competitors in the domestic and export markets.

A major aim of this chapter is therefore to analyse the structure of Kenya's manufacturing from the viewpoint of technical efficiency and productivity. These aspects are considered to crucial elements in establishing competitive strength in world markets.

8.3 Data and partial factor productivities

The RPED data contain extensive firm-specific information related to production, which opens up a number of possibilities for analysing various aspects of production. Since we use the production function as our main tool in characterising the production process within firms, we need to define and describe the outputs and inputs carefully. This is done next, followed by an analysis of partial factor productivities. Econometric estimation of production functions is conducted in sections 8.4 and 8.6.

The definitions of the output and inputs are as follows. Output (Y) is the value of all output produced by the firm in a given year. Capital (K) is the replacement value of the machinery and equipment. Labour (L) is the total number of full-time workers, including casuals, involved in production during the year. Intermediate inputs (M) is the sum of the costs for raw materials and energy. Some of the variables contain substantial shares of missing values, which have been imputed by utilising the panel dimension of the data set.[1] A small number of outliers are omitted from the analysis.

We expect the noise-to-signal ratio for the inputs and output to be comparatively high for a number of reasons. One is because of the problems associated with correctly measuring variables that vary over time, such as output and intermediate inputs. This problem is probably more serious for small firms, as argued by Liedholm and Mead (1987). Second, the value of output may be biased upwards due to varying degrees of market imperfections. A third reason is the difficulty of accurately valuing the capital stock, which typically consists of a wide range of different types of machinery and equipment acquired over long periods of time. A fourth potential cause for noise is unwillingness by the firms to reveal true values for fear this information reaches competitors or taxing authorities. A final source of data imprecision originates from the conversion of nominal to real values of the variables. Sector-specific deflators are used for outputs, while the deflators for capital and intermediate inputs are assumed equal across sectors. These deflators are described in detail in the Appendix. Given the heterogeneity of the sample firms, even within sub-sectors, one would have preferred the use of firm-specific deflators, but such indices were neither available from public sources nor easily deduced from the data.

Having mentioned these limitations, we may now consider the descriptive statistics of these variables presented in Table 8.1. A general observation for all variables is that their distributions are strongly skewed to the right. This is so despite the stratification of the sampling procedure in which larger and formal firms are overrepresented. The use of weights has a tremendous impact on the means: the weighted means, which are roughly representative of the 'true' population, are only a fraction of the sample means. The difference of gross output and intermediate input ($Y - M$), gross

Table 8.1 Summary statistics of output and input based on a sample of 615 (166 informal) observations on 266 firms (70 informal) in Kenyan manufacturing industries taken during 1993–95

	Mean	*Weighted Mean*	*Median*	*Std Dev.*	*Minimum*	*Maximum*
Y	45 632	2 109	5 400	113 811	4	1 150 000
K	47 130	1 099	3 500	167 556	1	2 500 000
L	102	6	22	301	1	4 000
M	24 543	1 165	1 916	68 366	0	905 250
$Y - M$	19 750	905	1 916	58 042	−43 409	764 940

Std Dev. = Standard Deviation.
Note: All variables except L measured in thousands of Kenyan shillings.

value added, is negative for 22 observations, which we interpret as indicative of inaccurate measurement primarily of the *M*-variable.[2]

Obscured in this table is the variation across firm categories. Naturally, the major difference occurs between formal and informal firms. For instance, the median number of workers among the sectors ranges from 40 to 51 in the formal sector in comparison with 2 to 5 in the informal sector. Among formal firms, the highest median output is found in the food sector, followed by metals. The largest median capital stock is recorded in metals, followed by food. Wood and textiles produce significantly less than these two sectors and operate smaller capital stocks.

Partial factor productivities and capital–labour ratios are presented in Table 8.2. They describe the contribution of a single input to output, which, in turn, may identify potential gains to be made from reallocating inputs to activities or firm types, in which they are more productive. Such ratios can also reveal differences in factor intensities, which *could* be evidence of imperfections in factor markets, such as price discrimination. However, these measures say nothing conclusive about total factor productivity, and neither do they inform us about a firm's technical efficiency.

Two measures of labour productivity are reported: the first is the simple output to the number of workers ratio (Y/L) and the second is the ratio of output to total wage cost (Y/wL). Both these measures display an inverted ∪-shaped relationship with firm size, peaking in the 76–500 and 21–75 worker intervals respectively. The increase in Y/L is more accentuated than the one in Y/wL, which is probably an effect of the higher wages paid by larger firms (see Chapter 4). Also, capital productivity (Y/K) and the capital–labour ratio reveal inverted ∪-shaped relationships. Another finding is that informal firms employ much less capital per employee than do formal firms. As a consequence capital productivity is higher, and labour productivity lower, for this category. Disaggregating the data into sub-sectors does not substantially alter these observations.

Table 8.2 Median relative factor productivities by size category for the pooled sample

	Y/L	Y/wL	Y/K	K/L
Firm category				
Informal	72	5.05	4.30	15
Formal, 1–5 workers	106	5.45	0.46	171
Formal, 6–20 workers	195	7.78	1.38	181
Formal, 21–75 workers	287	9.60	1.50	190
Formal, 76–500 workers	405	9.31	1.42	333
Formal, 500+ workers	187	8.63	0.82	260

Note: All variables except *L* are expressed in thousands of Kenyan shillings.

A short comparison with evidence from manufacturing sectors in other developing countries in Asia (Little, Mazumdar and Page, 1987) and Africa (Biggs, Shah and Srivastava, 1995) is instructive. Capital intensity generally increases with firm size, especially among African firms, although the smallest are not always the least capital intensive. Likewise, labour productivity increases with size in most cases, albeit not uniformly so, while capital productivity exhibits a less regular pattern, sometimes (in Thailand and South Korea) taking the form of the inverted-U detected above. In general terms, we may therefore conclude that our findings are broadly consistent with these studies, perhaps with the additional observation that the relationships in the Kenyan data are somewhat more stable.

As made clear above, one factor productivity that increases does not imply higher overall productivity, but if all factor productivities grow with firm size, it does. Indeed, this is the pattern we observe: firms in the 21–75, or possibly in the 76–500 worker interval, appear more productive. The next section will probe into the question of whether this is an effect of increasing returns to scale or a positive relationship between size and technical efficiency.

8.4 Productivity and technical efficiency

As discussed in the previous section, partial factor productivities are of limited usefulness in describing overall productivity of firms. Naturally, the concept of productivity refers to the ratio of output to input, which then by some index may be compared over time or across firms. The classical problem in productivity analysis is the case where there is more than one input: how should they be aggregated? A number of approaches exist, and the literature is rich in both theoretical contributions and empirical applications (see, for instance, Fried, Lovell and Schmidt, 1993). For our concerns, we have chosen the production function approach as the main tool of analysis. We justify this choice on both flexibility and transparency grounds: a simple production function may easily be complemented with explanatory variables in order to explore the determinants of productivity. The production function also provides parameter coefficients, which are directly interpretable, and usually accommodates statistical noise. The more sophisticated non-parametric approaches do not share these attributes, although they have other attractive properties.

Technical efficiency is a distinct concept, not to be confused with productivity, although they are closely related. The term refers to the ratio of actual output to the maximum output technically feasible given inputs. The set of maximum output levels corresponding to all positive input combinations forms the production frontier, the outer edge of the production possibility set, which in our case is defined by a frontier production function.

It follows that any pair-wise comparison of efficiencies among firms strictly depends on the shape of the estimated frontier. For example, there is nothing contradictory in a firm being both more productive and less efficient than another firm, which can happen if the firms differ in size and the technology exhibits non-constant returns to scale.

Hence, the scale property of the frontier production function is of great importance, especially if one wishes to make any statement about the size–efficiency relationship. As the estimates for returns to scale vary across different methodologies, it can sometimes be meaningful to impose the restriction of constant returns, a possibility considered below. Another important issue is on what level of aggregation firms are expected to share the same frontier production function. The approach taken in this section is to assume a single frontier for all sub-sectors, which carries the advantage of making intersectoral comparisons possible. Whether these differences really refer to technical efficiency can be questioned, however. In any case, they can certainly be interpreted as differences in productivity, which is the view taken below. The econometric models and the results are presented next.

8.4.1 Econometric models

The literature provides a number of methods for estimating technical efficiency. For our purposes, we have chosen to utilise two models: a fixed effects (FE) model and a random effects (RE) model, in order to check the invariance of our results to the particular method selected. Both models provide estimates of the production model:

$$y_{it} = \alpha + \mathbf{x}_{it}\beta + v_{it} - u_i \tag{8.1}$$

where y_{it} is output; α is an intercept term; x_{it} is a vector of inputs; β is a vector of coefficients; v_{it} is a disturbance term with errors assumed to be independently and identically distributed $N(0, \sigma_v^2)$; and $u_i \geqslant 0$ is a technical *in*efficiency term assumed fixed in the FE model and random in the RE model. The indices $i = 1, \ldots, N$ and $t = 1, \ldots, T$ refer to the firms and periods for which the observations were made, respectively. Motivated by the noisiness of the data and the short time span of the panel, we restrict technical efficiency to be time invariant.

In the FE model, we retrieve the u_is by first expressing (8.1) as a deviation from individual means

$$(y_{it} - \bar{y}_i) = (\mathbf{x}_{it} - \bar{\mathbf{x}}_i)\beta + (v_{it} - \bar{v}_i) \tag{8.2}$$

where the upper bar denotes individual means over time, and perform least squares on the resulting expression. This transformation removes the time invariant terms α and u_i. Second, we recover these parameters using,

$$\hat{\alpha} - \hat{u}_i = \hat{\alpha}_i = \bar{y}_i - \bar{x}_i\hat{\beta} \tag{8.3}$$

which is true because we assume the errors to average out to zero over time. In essence, this is the least squares dummy variable estimator. For a fully efficient firm, $u_i = 0$, and therefore $\hat{\alpha}_i = \hat{\alpha}$. Hence, we may finally compute the FE estimates of the u_is for the remaining firms using (Greene, 1993):

$$\hat{u}_i = \max(\hat{\alpha}_i) - \hat{\alpha}_i \tag{8.4}$$

This estimator is consistent in T, provided that also $\hat{\beta}$ is consistent, which holds as N goes to infinity. Throughout these estimations, we will use the natural logarithms of output and inputs. The appropriate measure of technical efficiency (*TE*) of firm i, in accordance with the original Farrell (1957) concept, is therefore:

$$TE_i = \exp(-\hat{u}_i) \tag{8.5}$$

which is 1.00 for fully efficient firms and in the (0, 1) interval for firms with some degree of technical inefficiency of production.

In the RE model, we treat the u_is in equation (8.1) as random, and we will assume that they are independent from the v_{it}s and iid $|N(0,\_\sigma_u^2)|$ (which means that its distribution is truncated from below at zero to ensure non-negativity). This model is a special case of the model proposed by Battese and Coelli (1992) with the efficiency time trend parameter and the parameter for the mean of the truncated distribution of u_i restricted to zero. The parameters in (8.1) and the variance parameters, $\sigma^2 = \sigma_u^2 + \sigma_v^2$ and $\gamma = \sigma_u^2/(\sigma_u^2 + \sigma_v^2)$, are estimated using the method of maximum likelihood utilising the computer programme, FRONTIER 4.1b (Coelli, 1994). The significance of the inefficiency effects can be tested using a likelihood ratio test of the γ-parameter. Firm-specific technical efficiencies are predicted by deriving expressions for the conditional expectation of the right-hand side in equation (8.5) conditional upon the observed value of $(v_{it} - u_i)$ in the manner set forth by Battese and Coelli (1993).

The issue of whether u_i should be treated as fixed or random is by no means trivial. Both approaches have strengths and weaknesses. The FE model permits the firm-specific effects to be correlated with the explanatory variables: the parameter estimates are still unbiased and consistent. Moreover, the errors need not be normal and if heteroskedasticity is present, the estimates remain consistent, albeit inefficient. On the other hand, the approach is very sensitive to statistical noise. A single erroneous observation may have profound effects on the obtained efficiency scores. The RE model explicitly accommodates the noise component, but produces biased and inconsistent estimates if the errors are not independent from the inputs. Furthermore, the predicted efficiencies of the RE model have been shown to be very sensitive to heteroskedasticity in any of the two error terms (Caudill and Ford, 1993). In light of this discussion, it is evident that restricting attention to only one methodology is potentially hazardous. We shall therefore consider both, and it is shown below that the main findings are robust as regards the choice of approach.

An additional difficulty concerns the potential endogeneity of inputs. If an input in (8.1) is endogenous, in the sense that it is a function of output, the estimated parameters are biased and inconsistent. The conventional way to solve this problem is to apply some type of Instrumental Variable-estimation technique, but for production functions appropriate instruments are rarely available. In a panel data setting, a natural instrument is lagged inputs, but it is doubtful whether such instruments are econometrically valid, especially so in our case because of the simultaneous presence of both growing and contracting firms.[3] Earlier experience also suggests a momentous drop in statistical efficiency by using this approach. Consequently, we have decided to estimate the production functions directly, acknowledging the fact that the parameter estimates may be subject to some unknown degree of simultaneity bias.

8.4.2 Results

The parameter estimates obtained from estimation of the FE and the RE models outlined above are presented in Table 8.3. The production functions are restricted to the Cobb-Douglas form, motivated by its small number of parameters and the poor performance of the more flexible translog specification regarding compliance with the regularity conditions. Given our doubts about the accuracy of the *M*-variable, we specified *Y* as a function of *K*, *L* and *t* only, simply omitting that variable. Alternative specifications, such as $Y - M$ as a function of *K*, *L* and *t*, or *Y* as a function of *K*, *L*, *M* and *t*, do not change the main conclusions drawn.

The FE models are estimated using weighted OLS to correct for heteroscedasticity, not for the stratification of the sample.[4] Only pooled results are reported in order to make intersectoral productivity comparisons possible, as mentioned above. Thus, we confine all firms to share the same technology, which is defensible because none of the estimated coefficients of the pooled model is significantly different from the ones obtained from the sub-sectors estimated individually. The only significant difference is the time variable in textiles, which is negative and significant at the 5 per cent level. We return to the interpretation of this result later.

The parameter estimates for the inputs in Table 8.3 are all significant and interpretable as input elasticities with respect to output because of the

Table 8.3 Parameter estimates of the fixed effects model (FE), the fixed-effects model with constant returns to scale imposed (FE-CRS), and the random effects model (RE)

	FE *Coefficient*	*FE-CRS* *Coefficient*	*RE* *Coefficient*
Intercept	0.001(0.017)	0.002(0.017)	9.718(0.329)
ln (*K*)	0.108(0.031)	0.168(0.031)	0.278(0.029)
ln (*L*)	0.504(0.063)	0.832(0.031)	0.839(0.056)
t	−0.037(0.022)	−0.021(0.022)	−0.026(0.042)
σ_{-}^{2}			2.298(0.255)
γ			0.715(0.042)
Returns to scale	0.612	1.000	1.117
R-square	0.150	0.095	
Log-likelihood			−878.540
LR-test*			93.910
Observations†	555	555	615
Tests (Prob values):			
CRS††	0.000		0.005
Homoscedasticity§	0.342	0.266	

Standard errors in parentheses.
*This is a likelihood ratio test of the significance of γ. Critical value at the 5% level with one degree of freedom is 2.71 (obtained from the mixed chi-square distribution in Table 1 in Kodde and Palm, 1986).
†The within transformation removes all firms that were only observed once. Hence, the number of observations is lower in the FE-models (555 compared to 615).
††Wald test of the linear restriction, $\ln(K) + \ln(L) = 1.00$.
§White's test for homoscedasticity.

Cobb-Douglas functional form. The returns to scale, the sum of the input elasticities, of the unrestricted FE model in the second column is remarkably low and significantly less than unity as indicated by the Wald test reported at the bottom of the column. Similar estimates on returns to scale are obtained by Roberts and Tybout (1996, Table 5.3) who consider them too low to be plausible, and probably driven by sample noise.[5] We have therefore estimated a second FE model imposing the restriction of constant returns to scale (FE-CRS).[6] In the RE model, we find significantly increasing returns to scale: in comparison with the FE-CRS model this is mainly driven by a significantly higher elasticity for capital.

Mean technical efficiencies are reported in Table 8.4. For the FE models these were calculated by replacing max $(\hat{\alpha}_i)$ with the ninetieth percentile, in the process defining the upper decentile (in terms of $(\hat{\alpha}_i)$) of firms as technically efficient. This is motivated by a desire to make the frontier 'less deterministic' because of the paramount influence a single large positive residual may have on overall efficiency scores. The use of the ninetieth percentile is certainly arbitrary, but all conclusions based on comparisons

Table 8.4 Mean predicted technical efficiencies by firm category for three different models

Firm category	*Obs (FE)*	*FE*	*FE-CRS*	*Obs (RE)*	*RE*
Informal	56	0.04	0.20	70	0.41
Formal	155	0.34	0.43	196	0.45
Formal, 1–5 workers	19	0.09	0.24	22	0.35
Formal, 6–20 workers	35	0.20	0.39	44	0.45
Formal, 21–75 workers	59	0.34	0.45	73	0.47
Formal, 75–149 workers	23	0.50	0.52	29	0.49
Formal, 150+ workers	19	0.68	0.51	28	0.42
Food, formal	41	0.49	0.60	56	0.54
Wood, formal	39	0.23	0.31	44	0.38
Textiles, formal	34	0.21	0.32	49	0.40
Metals, formal	41	0.41	0.47	47	0.47
All	211	0.26	0.37	266	0.44

The number of observations in each cell for the fixed-effect models (Obs(FE)) and the random effects model (Obs (RE)) are reported for each cell.

between firm categories remain irrespective of whether the eightieth, ninety-fifth or ninety-ninth percentile is estimated. The only change is a shift in the overall *level* of mean efficiencies. For this reason quantitative comparisons among firms in terms of efficiency units are somewhat precarious, whereas the qualitative differences are more certain. Given the similarity of mean efficiencies to those of the stochastic approach, however, we believe that the predicted efficiencies by the FE models are roughly correct.

The cell means across firm categories reveal several interesting relationships. All models predict strongly positive size–efficiency relationships up to and including the 75–149 worker segment. For the largest size category, efficiency declines in the FE-CRS and RE models, proposing a similar inverted ∪-shaped relationship as discovered in the previous section. This pattern does not occur in the FE model, which we regard as a less likely result, probably driven by the unrealistically low returns to scale.

Informal firms are less efficient than formal ones in all models, although the magnitude of this difference varies considerably across models. The higher the estimated returns to scale, the smaller is this difference. In the FE and FE-CRS models, formal firms are more than twice as efficient as their informal competitors, whereas the difference is less than 10 per cent according to the RE model. The small sizes and low capital–labour ratios of the informal firms probably cause this variation (see Table 8.2), which locates them at the far corner of the production frontier hyperspace.

As argued above, comparisons across sectors should be interpreted as differences in productivity, not efficiency, because of the possibility that the intercept of the frontier production function may differ across sectors. We may thus conclude that food is the most productive sector, followed by metals. Productivity in food is 40–120 per cent higher, and in metals 20–85 per cent higher, than in wood and textiles, depending on model. These relationships hold when size categories are compared across sectors.

8.5 Review of potential determinants of productivity

In the previous section, we analysed the variation in technical efficiency and productivity across sectors and categories of firms. A logical second step is then to unveil the underlying causes for these differences. Potential determinants are scrutinised below, together with their expected influence on performance. Empirical estimations are conducted in section 8.6 using average production functions augmented with the determinants. Our analysis therefore focuses on productivity, not technical efficiency, although the arguments presented below relate to both concepts. We prefer this approach to an analysis of the determinants for technical efficiency because it is simpler, does not rely upon the assumption that efficiency is independent of inputs, and that parameter coefficients of all variables are directly interpretable in a straightforward manner. Also, if the background variables of interest for the performance of manufacturers are available a priori, as is the case here, OLS regression will 'do just fine', as Knox Lovell (1993, p. 7) put it.

For natural reasons, the discussion below focuses on firm-specific factors rather than factors that are common to all firms. Of course, factors such as corruption, regulations, taxes, infrastructure, overall education levels and cultural attitudes by the workforce, aggregate demand, political instability and institutional efficiency all affect performance of firms to some extent. But the point is that they affect all firms more or less equally, and they do not predict the variation between firms in any foreseeable way. The firm-specific factors we consider are: location, age, growth, investment rate, capacity utilisation, skilled labour ratio, race of owners, whether or not the firm exports, degree of foreign ownership, and access to credit.

The location variable considers the different effect on productivity that may exist between the three categories of towns. Mombasa, the main port of entry, would confer transportation cost advantages on both imported inputs and exports. Nairobi has the advantage of being the largest city with a more developed infrastructure relative to the smaller cities of Eldoret and Nakuru. Eldoret, the newest and fastest-growing industrial town, could also enjoy some advantages.

The inclusion of the age and growth variables is suggested by the literature on firm growth. A seminal contribution to this literature is a selection model by Jovanovic (1982), in which the underlying cause for different growth patterns among firms is the variation in their efficiencies. Profit-maximising output is an increasing function of firm-level efficiency that is fixed. Since the efficiencies are unknown to the managers, they must infer it through operating in the market, a process that takes time because of noise. Over time, efficient firms grow and survive, while inefficient firms remain small or exit. This implies that larger firms are more efficient than small, and that growing enterprises are more efficient than contracting ones. Also, older firms, on average and for a given age, are more efficient than younger ones as a result of the selection process taking place. Several variants of this model exist (see, for instance, Hopenhayn, 1992, Pakes and Ericsson, 1995) but its main results hold. If the Kenyan manufacturing firms conform to this theory, we expect both the age and growth variables to have positive signs. In addition, a positive age variable may also be an effect of learning-by-doing, leading to accumulation of experience and improved productivity over time.

The investment rate is ambiguous. If it proxies firm growth, we may expect it to be positive because of the argument above. On the other hand, if adjustment costs of capital investment are present, it can be negative. It is notable that the investment theory literature incorporates neither efficiency nor productivity explicitly in the analysis (Chirinko, 1993). The degree of capacity utilisation must have a positive effect on productivity unless the capital stock has reached levels of congestion, which is not likely anywhere in Africa. Human capital is captured in the ratio of skilled labour to the total labour force, and is expected to have a positive effect on productivity. Its relative impact can be regarded as a test of Hamermesh's (1993) claim that the main constraint in many developing countries is skill, not physical capital. The race factor aims to isolate the relative inexperience of the African late starters and hence their relatively shorter duration of learning-by-doing. The Africans' relative lack of access to financial and other related resources works in the same direction.

The export promotion hypothesis postulates that exporting firms are more productive than non-exporters. Productive firms self-select into exporting. The competitive pressure in the export markets can also spur firms to raise their performance. The degree of foreign ownership may have a positive influence if it involves better technical support to the firm, or tighter monitoring of managers to be efficient, had the owners been exclusively Kenyan. If working capital is a main constraint for firms, we may expect access to credit to contribute positively to productivity. As informal firms are distinct in terms of low age, no access to formal credit, high degree of African ownership, no export involvement or foreign ownership, inclusion of an informal dummy would lead to severe multicollinearity problems among the explanatory variables in the following analysis. It is therefore omitted.

Naturally, the direction of causality of these determinants is not always obvious. In fact, for some it may go both ways. For example, are firms more productive because they export, have access to credit and have foreign owners, or is it the other way around, so that they possess all these characteristics because they are productive? Or can both statements be true? Clearly, one can certainly argue for either case. However, it is beyond the scope of this chapter to settle this issue and we shall therefore regard the estimated relationships in the next section as associations rather than casual links.

The definitions of the determinants are as follows. Firm age (*Age*) is measured in years since operations started. The growth variable (*Growth*) is defined as, $Growth = \Delta L_t / \max(L_{t-1}, L_t)$, which is bounded in the $[-1, 1]$ interval. The investment variable is total investment in capital goods undertaken during the year divided by *K*.[7] The capacity variable (*Capacity*) is defined as the utilisation rate of the capital stock under current working hours. The skill variable is defined as the ratio of skilled to the total number of workers.[8] Because of measurement problems associated with this variable for small firms, it is set to zero for establishments with fewer than ten workers. The *African* and *Export* variables are dummies for African ownership and involvement in export markets, respectively. Foreign ownership (*Foreign*) is measured in percentages, and the *Credit* variable is a dummy variable taking the value of one if the company has access to an overdraft facility at a bank.

Means of these determinants are presented in Table 8.5. In terms of age, capacity and skill, the sectors are notably similar. Considering other variables, some interesting differences emerge. It is intriguing that the growth rates follow the ranking in efficiency levels in section 8.4, with food at the top, followed by metals. These two sectors are also much more involved in export markets than wood and textiles.

The bivariate Pearson rank correlations in Table 8.6 display a number of significant associations among the variables. Perhaps the strongest correlation is with firm size, measured as the natural logarithm of the number of workers, which is positively related to age, skill, export, foreign ownership and credit.

Table 8.5 Sample means of the nine productivity determinants

	All	*Food*	*Wood*	*Textile*	*Metal*
Age	19.86	20.52	20.94	20.18	17.96
Growth	−0.01	0.02	−0.04	−0.05	0.00
Investment	0.08	0.10	0.08	0.05	0.07
Capacity	0.64	0.65	0.68	0.62	0.61
Skill	0.20	0.18	0.22	0.20	0.19
African	0.37	0.29	0.37	0.34	0.48
Export	0.23	0.27	0.16	0.19	0.31
Foreign	9.76	10.26	12.52	6.14	9.73
Credit	0.60	0.75	0.56	0.56	0.54

Table 8.6 Correlation matrix of the productivity determinants

	Age	*Growth*	*Investment*	*Capacity*	*Skill*	*African*	*Export*	*Foreign*	*Credit*
Growth	−0.17*								
Investment	−0.22*	0.16*							
Capacity	0.04	0.09*	0.08*						
Skill	0.24*	−0.06	−0.17*	0.07**					
African	−0.40*	0.04	0.14*	−0.10*	−0.34*				
Export	0.14*	0.04	0.03	0.05	0.19*	−0.27*			
Foreign	0.25*	−0.03	−0.05	0.05	0.08*	−0.27*	0.34*		
Credit	0.33*	−0.09*	−0.15*	0.04	0.39*	−0.50*	0.36*	0.22*	
Size	0.38*	0.00	−0.08*	0.21*	0.42*	−0.55*	0.57*	0.33*	0.65*

Note: * and ** signify significance at the 5 per cent and 10 per cent confidence levels, respectively.

Likewise, export is positively correlated with age, foreign ownership and credit. The growth variable has, surprisingly, the smallest and least significant coefficients. African ownership is strongly correlated with young age, lack of skill, little export involvement, small degree of foreign ownership, rare access to formal credit and small size. As argued above, these are the characteristics of informal firms, which motivated the exclusion of an informal dummy variable from the productivity analysis below on multicollinearity grounds.

8.6 The determinants of productivity

The analysis of the determinants for productivity is conducted using OLS regression with output as the dependent variable, and the inputs and determinants as explanatory variables. As with the FE models above, weights are used in the estimations to correct for heteroscedasticity. For some of the determinants, as discussed in the previous section, the direction of causality may go both ways. For this reason, the estimated parameters should be interpreted as the nature of *association*, rather than *causation*, between the determinants and productivity. It follows that the parameter estimates may be subject to the same unknown degree of simultaneity bias, as discussed in section 8.4.

To explore some of the heterogeneity of the sample, we present both pooled and sector models in Table 8.7. The labour elasticities are somewhat higher than estimated with the models in section 8.4, whereas the capital elasticities are in between those estimated with the FE and RE models. Constant returns to scale cannot be rejected in any model except in textiles according to the Wald tests reported at the bottom end of the table, which is consistent with our previous analysis. The time variable is insignificantly different from zero in all models, except textiles, which means that no technical change is picked up in the regressions.

Thus, textiles behave rather differently compared to the other sectors as both significant decreasing returns to scale (0.6) and negative technical change (−0.2) is estimated in the sector. Most probably this is an effect of

Table 8.7 Parameter estimates of productivity regressions (pooled and sector models reported)

	Pooled Coefficient	*Food Coefficient*	*Wood Coefficient*	*Textiles Coefficient*	*Metal Coefficient*
Intercept	9.118*	8.772*	9.557*	13.138*	9.077*
	(0.461)	(1.110)	(0.872)	(1.039)	(1.025)
ln(*K*)	0.249*	0.246*	0.196*	0.191*	0.255*
	(0.030)	(0.083)	(0.060)	(0.068)	(0.064)
ln(*L*)	0.733*	0.749*	0.824*	0.422*	0.788*
	(0.051)	(0.134)	(0.104)	(0.114)	(0.103)
t	−0.010	0.030	0.017	−0.228*	0.128
	(0.046)	(0.122)	(0.089)	(0.088)	(0.092)
WOOD	−0.677*				
	(0.120)				
TEXTILES	−0.784*				
	(0.127)				
METAL	−0.435*				
	(0.118)				
NAI	0.494*	0.135	0.630*	−0.480**	0.215
	(0.108)	(0.247)	(0.184)	(0.263)	(0.306)
MOM	0.459*	0.513	0.261	−1.320*	0.448
	(0.142)	(0.357)	(0.309)	(0.300)	(0.354)
ln(*Age*)	−0.039	−0.003	−0.196	−0.372*	−0.265
	(0.065)	(0.165)	(0.130)	(0.122)	(0.183)
Dgrowth	−0.216*	0.085	−0.230	−0.185	−0.281
	(0.100)	(0.262)	(0.214)	(0.197)	(0.188)
Dcontract	−0.234*	−0.500*	−0.099	−0.138	−0.444*
	(0.087)	(0.250)	(0.163)	(0.158)	(0.206)
Investment	0.444	0.144	0.183	0.250	1.111**
	(0.312)	(0.632)	(0.553)	(0.772)	(0.671)
Capacity	0.363*	0.135	0.323	0.857*	0.339
	(0.165)	(0.421)	(0.339)	(0.289)	(0.334)
Skill	0.254	1.223*	−0.032	0.111	−0.203
	(0.162)	(0.522)	(0.280)	(0.270)	(0.421)
African	−0.219*	−0.209	−0.329	−2.334*	−0.262
	(0.109)	(0.246)	(0.217)	(0.351)	(0.265)
Export	0.228*	0.330	0.521*	0.135	−0.156
	(0.103)	(0.248)	(0.223)	(0.202)	(0.209)
Foreign	−0.001	0.001	−0.004**	0.008**	0.004
	(0.001)	(0.004)	(0.002)	(0.004)	(0.003)
Credit	0.503*	0.902*	0.218	−0.044	0.677*
	(0.110)	(0.293)	(0.187)	(0.219)	(0.286)
Observations	602	114	152	127	142
R-square	0.861	0.855	0.854	0.904	0.893
White test†	0.333	0.746	0.549	0.737	0.499
CRS test††	0.630	0.952	0.820	0.000	0.596

* and ** signify significance at the 5% and 10% confidence level, respectively.
† White test of heteroscedasticity. The null is homoscedasticity. Probability value reported.
†† Wald test of the linear hypothesis: $\ln(L) + \ln(K) = 1.00$. Probability value reported.

the dramatic drop in aggregate output of almost 50 per cent in the sector following the sudden inflow of imported used clothes, which occurred in the middle of the sampling period. As a response to this knife-edge competition, many producers were unable to survive the falling prices and began to reorient their activities towards alternative areas, such as services. A general impression from the field by the interviewers was that such transformations were a lot easier for small than for large firms. The low estimate for returns to scale may therefore capture the inability of the larger establishments, with their specialised production lines and machinery, to shift market niche.

The sector dummy variables in the pooled model confirm our earlier findings that food, followed by metal, is the most productive sector. The coefficients suggest that food is approximately 70 per cent more productive than wood and textiles, and 40 per cent more productive than metals, which is roughly consistent with our earlier intersectoral comparisons. The Nairobi and Mombasa dummy variables are significantly positive in the pooled model, suggesting that location in the capital or in the main trading port may influence productivity positively. The location influence is positive in all sector models, excluding textiles, but is generally insignificant. Interestingly, textile firms, located especially in Mombasa but also in Nairobi, are significantly less productive than those based in Eldoret and Nakuru. Presumably, the negative consequences of competition discussed above are stronger in places where this competition was most serious. Mombasa, the principal port for overseas imports, is certainly such a place as no additional transport costs inflated the price of imported clothes.

The firm age variable is insignificant everywhere except textiles. Hence we find no support for the Roberts and Tybout (1996) argument that productivity increases with firm age until a threshold age of about four years, and neither does it support the hypothesised positive age–productivity relationship in the firm growth literature. The textiles exception may be due to older firms also being the larger ones, which were struck particularly hard by the trade liberalisation mentioned above.

The firm growth variable was divided into two separate variables to allow for distinct growth–productivity relationships for expanding and contracting firms. Because of the poor quality of this variable in terms of magnitude,[9] but not in sign, we entered firm growth in the regressions as dummy variables. Specifically, we defined the variable *Dgrowth* to takes the value of one for firms with a positive growth rate of 3 per cent or more, and zero otherwise, and the variable *Dcontract* to take the value of one for firms with a negative growth rate exceeding 3 per cent, and zero otherwise. Interestingly, *Dgrowth* is negative and significant in the pooled model, which contradicts our expectations, but not so in the sector models. On the other hand, *Dcontract* is significantly negative in the pooled, food and metal models, which is consistent with most theories of firm growth, including the Jovanovic model, and with our discussion in the previous section.

The investment ratio is positive but insignificant everywhere, except metals where the variable is significant and positive at the 10 per cent level. So, neither adjustment costs nor selection effects, in which the more productive firms undertake investment, seem to be particularly traceable in the data. Capacity enters with the predicted positive sign, and is significant in the pooled and textiles models. The human capital variable does not influence productivity significantly in any sector except food, where the effect is positive and rather strong.

Enterprises with African owners appear less productive in all models. The effect is significant in the pooled and textiles models. It is revealing to compare this finding with the estimated effect of the *Cred* variable, which is positive and significant in all models except wood and textiles. In fact, informal firms, almost without exception, lack this facility. Also, African owners are much more common in the informal than in the formal sector. Together, these two variables confirm our previous finding of considerably lower performance of informal firms.

The export variable is positive and significant in the pooled and wood models, but the magnitude and significance is less than expected and does not lend very strong support for the efficiency arguments of export promotion. The impact of foreign ownership plays a minor role, and takes both negative and positive signs for the various models, but is insignificant at the 5 per cent level in all estimations.

Taken together, the results presented above have demonstrated a substantial amount of heterogeneity across sectors.

8.7 Summary and conclusions

We have in this chapter analysed the performance of the manufacturing sectors in terms of technical efficiency, technical change and productivity. Our basic tool of analysis was the production function, which served as a best-practice frontier in the measurement of efficiency, and as an average response function in the analysis of the determinants of productivity.

It was found that the estimated input elasticities and returns to scale are strongly dependent on methodology. The results from the unrestricted fixed-effects model suggest decreasing returns, whereas the stochastic frontier model and the average production functions propose increasing and constant returns, respectively. We did not try to distinguish among these models, but we note that this finding is consistent with Roberts and Tybout (1996) and supports their claim that correctly measuring scale returns can be difficult with noisy data. The mean efficiencies vary across models (see Table 8.4); however, the rankings of sectors and firm categories are basically the same for all models.

We found a strong positive size–efficiency relationship. The FE model with constant returns imposed suggests formal firms in the 75–149 worker category to be about twice as efficient as formal firms in the 1–5 worker interval. There is weak evidence that efficiency is somewhat lower for establishments in the 150+ worker category, indicating the possible presence of an inverted ∪-shaped size-efficiency relationship. Such a relationship is evident, and much clearer, with the labour and capital productivities which peak in the 25–74 and 75–500 worker categories, respectively. When comparing productivity across sectors, it was found that food is 40–120 per cent, and metal 20–85 per cent, more productive than wood and textiles, depending on model.

In all models, textiles behaved differently from the other sectors when estimated individually. Strongly decreasing returns, a negative and significant time trend and negative influence of location in Mombasa separate it from the other sectors. We believe these results are driven by heavy costs of adjustments following the inflow of used clothes in the middle of the sample period in which gross sector output dropped by almost a half. The general observation from the field that larger firms were less able to adjust quickly may contribute to the estimated decreasing returns.

The analysis of the determinants for productivity revealed some heterogeneity among sectors, which makes it precarious to draw general conclusions in some cases. Apart from textiles, the location variables indicate positive impacts on productivity by operating in Mombasa or Nairobi, although this effect is weak and significant only in the pooled model. The age variable fails to explain much of the variation in productivity, as do the dummy variables for growing firms and the investment ratio. However, contracting firms are shown to be less productive in all models, which is consistent with findings from other developing countries in Roberts and Tybout (1997). In line with their reasoning, we may therefore conclude that an increase in the turnover rate of enterprises in Kenyan manufacturing, in which contracting firms cease operations, is likely to raise overall sector productivity. Naturally, higher turnover also implies lower mean age of firms, but since age has no effect this does not affect overall sector performance according to our results.

Firms with African owners and limited access to credit are less productive, and because these are the main characteristics of informal firms, it appears as if productivity is lower in the informal sector. The skill variable does not confirm the claim that human capital is the main constraint for production in developing countries: it is positive in food but insignificant everywhere else. Further, we find weak support for the export penetration hypothesis, that exporting implies higher productivity, and no support for the argument that foreign ownership has an influence.

Based on these findings, we believe that three conclusions relevant for policy can be drawn. The first concerns the dramatic effects imposed on textiles caused by the import liberalisation of cheap used clothes. The possible gains in consumer surplus must be weighed against the heavy adjustment costs incurred by the textiles sector, and we suspect that a more gradual reform might have been warranted. A second conclusion is based on the observation that overall sector productivity may increase with higher enterprise turnover. To encourage such a process, policy needs to consider measures that stimulate births of new firms, and not those that generally support existing firms. A third conclusion is that access to credit appears to be of relatively more importance than export involvement and human capital. This suggests that the government should prioritise reforms that improve access to credit by private firms.

Appendix

Deflators

The output deflators were calculated as $p_t = (pq)_t / q_t$ ($t = 1992, 1993, 1994$), where $(pq)_t$ is output by sector, and q_t is the quantity index (Table 94c and Table 92 respectively, Central Bureau of Statistics, 1995b). In the food sector, two output deflators were used, one for bakeries and another for 'total food manufacturing' to account for the unique drop in price for bakery products in 1994. The capital deflator is calculated using data on capital formation of private machines and equipment in the manufacturing sector which are reported in current and fixed prices in Central Bureau of Statistics (1995a), Table 43a–b. Intermediate inputs is deflated using the output deflators.

Table A8.1 Output and input deflators for Kenyan manufacturing sectors

Deflators, 1992 = 100	*Year*		
	1992	*1993*	*1994*
Output deflators by sector:			
Food (total food manufacturing)	100	135	166
Food (bakery products)	100	137	109
Wood (wood and cork products)	100	122	146
Textiles	100	127	209
Metal	100	132	145
Capital deflator	100	123	165
CPI deflator (for comparison, not used)	100	146	188

Notes

1. Missing entries of capital are imputed using previous or last period's value, correcting for the depreciation rate. Each intermediate inputs is imputed by using its mean ratio to output in the periods for which the variable is reported. Output and labour are not imputed.
2. Y is a single figure, whereas M is a summary measure of all intermediate inputs, some of which the respondent had problems recalling accurately. For this reason, we believe that Y is measured with more precision than M.
3. Thus, predicted inputs would exceed true values for contracting firms and therefore underestimate their true efficiencies, and vice versa for growing firms.
4. Because all firms are assumed to share the same technology, defined by a production function, weighting the sample according to sampling probability would produce the same results asymptotically. If different strata of the sample have different slope parameters, should separate production functions for these be

estimated (Deaton, 1997)? Further, giving the small firms (for which the error variance is greater) larger weight would increase the problem of heteroscedasticity, which in the presented setting is solved by increasing the weights for the larger firms.

5. Deaton (1997, p. 108f) shows that measurement error in the explanatory variables constitutes a bias towards zero of within slope parameter estimates.
6. We would certainly have favoured estimating the RE model with the same restriction, but this is not easily implementable with the current state of software.
7. Values exceeding 0.5 are considered implausible and therefore set to 0.5. The correction is made for 31 observations.
8. Skilled workers include managers, proprietors, engineers, physical scientists, accountants, economists, technicians, foremen, supervisors, and specifically skilled production workers.
9. The growth variable is not based on the input *L*, but on related data on net hiring during the year because *L* was not observed for the previous period for half of the observations. The two variables, where observed, differ in size but rarely in sign, which motivates the use of growth dummy variables.

References

Battese, G. E. and Coelli, T. J. (1992), 'Frontier production functions, technical efficiency and panel data: With application to paddy farmers in India', *The Journal of Productivity Analysis* 3.

Battese, G. E. and Coelli, T. J. (1993), 'A stochastic frontier production function' Working paper in econometrics and applied statistics No. 69, Department of Econometrics, University of New England, Armidale.

Biggs, T., Shah, M. and Srivastava, P. (1995), *Training, Technology and Firm Efficiency in Sub-Saharan Africa*, RPED case study series, World Bank, Washington, DC.

Caudill, S. B. and Ford, J. M. (1993), 'Biases in frontier estimation due to heteroscedasticity', *Economic Letters* 41.

Central Bureau of Statistics (1995a), *Economic Survey 1995*, Office of the Vice-President and Ministry of Planning and National Development, Nairobi.

Central Bureau of Statistics (1995b), *Statistical Abstract 1995*, Office of the Vice-President and Ministry of Planning and National Development, Nairobi.

Chirinko, R. S. (1993), 'Business fixed investment spending: Modelling strategies, empirical results, and policy implications', *Journal of Economic Literature*, 31(4).

Coelli, T. (1994), *A guide to FRONTIER version 4.1: A computer program for stochastic frontier production and cost function estimation*, Department of Econometrics, University of New England, Armidale.

Deaton, A. (1997), *The Analysis of Household Surveys*, The John Hopkins University Press, London.

Farrell, M. J. (1957), 'The measurement of productive efficiency', *Journal of the Royal Statistical Society* 120.

Fried, H. O., Lowell, C. A. K. and Schmidt, S. S. (eds) (1993), *The Measurement of Productive Efficiency*, Oxford University Press, New York.

Greene, W. H. (1993), 'The econometric analysis to efficiency analysis', in Fried Lowell and Schmidt (1993), pp. 68–119.

Hamermesh, D. S. (1993), *Labour Demand*, Princeton University Press, Princeton, NJ.

Hopenhayn, H. A. (1992), 'Entry, exit and firm dynamics in long run equilibrium', *Econometrica* 60.

Jovanovic, B. (1982), 'Selection and the evolution of industries', *Econometrica* 50.
Knox Lovell, C. A. (1993), 'Production frontiers and productive efficiency', in Fried, Lowell and Schmidt (1993), pp. 3–67.
Kodde, D. A. and Palm, F. C. (1986), 'Wald criteria for jointly testing equality and inequality restrictions', *Econometrica* 54.
Liedholm, C. and Mead, D. (1987), *Small Scale Industries in Developing Countries: Empirical Evidence and Policy Implications*, USAID/Dept of Agricultural Economics, Michigan State University, East Lansing, MI.
Little, I., M., Mazumdar, D. and Page, J. M. (1987), *Small Manufacturing Firms. A Comparative Analysis of India and Other Economies*, Oxford University Press, for the World Bank, London.
Pakes, A. and Ericson, R. (1995), 'Markov-Perfect Industry Dynamics: A Framework for Empirical Work', *Review of Economic Studies* 62(1).
Roberts, M. J. and Tybout, J. R. (eds) (1996), *Industrial Evolution in Developing Countries: Micro Patterns of Turnover, Productivity, and Market Structure*, Oxford University Press, London.
Roberts, M. J. and Tybout, J. R. (1997), 'Producer Turnover and Productivity Growth in Developing Countries', *The World Bank Research Observer* 12(1).

9
Export Performance in the Kenyan Manufacturing Sector

Mats Granér and Anders Isaksson[1]

9.1 Introduction

Manufacturing for exports has become a major element in strategies for industrial growth in the developing world. The successful experience of East Asian countries has underscored the importance of manufactured exports in strategies for expanding industrial production and for overall growth. Recently, sub-Saharan African countries have also embarked on this strategy, although many of these countries still rely on primary exports. Kenya is one of the African countries that has adopted a policy of manufacturing export promotion.

The policy of outward orientation has a number of attractive features. For instance, increased international competition may enhance the productivity of firms and thus promote growth. Furthermore, as the market grows economies of scale may be exploited and there may be important domestic spillover effects. These potential benefits are often particularly associated with manufactured exports. Moreover, the increased trade diversification emanating from manufacturing exports has a stabilising effect and earnings from manufactured exports offer better support for stable growth compared with primary exports (Helleiner, 1995). It is also likely that macroeconomic policy is improved under conditions of export-orientation. In addition, empirical evidence seems to support the view that trade liberalisation and trade openness are closely correlated with GDP growth.[2]

From a policy-making point of view, it is, therefore, important to understand what determines the firms' decision to export. It is also vital to know, given that a firm has decided to export, the determinants of how much of output is exported. If these determinants can be identified, policy can be directed towards improving the conditions of potential exporters. These determinants are the main concerns of this chapter.

To shed some light on these issues, we use data on Kenyan manufacturing firms collected between 1993 and 1995. To date, there exist few studies investigating manufacturing exports using firm-level data on sub-Saharan

African countries.[3] Interestingly, the time period covered in the data was characterised by a move towards greater openness, as evidenced by the removal of quotas, lowering and unification of tariffs, and an active depreciation policy to make the exchange rate more realistic and competitive. We therefore believe that information related to firms' response to trade liberalisation in general, and export promotion in particular, can be found in the data.

The chapter is organised as follows: section 9.2 describes the theoretical link between increased exports and GDP growth. We then turn to possible hypotheses concerning the decision to export and the amount of exports. In section 9.3, we profile the Kenyan trade liberalisation and provide some macro-data on manufacturing sector trade performance. In section 9.4, we look at micro-evidence of export performance in relation to various firm characteristics. Econometric analysis of the data is undertaken in section 9.5, while section 9.6 concludes this chapter.

9.2 Exports and growth

In this section we first present the neoclassical arguments for the export-promotion policy. We also provide some complications to this theory. Thereafter, we discuss the two main determinants, at least in theory (namely sunk costs and future profitability of exporting) behind the decision to export. We also discuss the theoretical determinants of how much of output to export.

9.2.1 Export-led growth

An important part of the structural adjustment programmes implemented in many African countries has been trade liberalisation, where the main objective has been to increase manufacturing exports. The idea is that increased exports enhance economic growth. For this to happen, the number of exporters and/or the exports from existing exporters must increase.

Several explanations for the causal link between exports and productivity can be put forward. First, exports tend to concentrate investment in the most efficient sectors in the economy: those in which the country enjoys a comparative advantage. Stronger specialisation in these sectors increases productivity. Second, access to foreign markets allows a country to gain from economies of scale as the domestic market is enlarged by the inclusion of an international component, thus permitting larger-scale operations than were previously possible. Third, stronger exposure to international competition is considered to increase the pressure on the export industries to keep costs low and to provide incentives to reduce technical inefficiency. This results in productivity gains. Fourth, exporting firms can reduce technical inefficiency through learning-by-exporting effects stemming from the recipients' technical expertise. Fifth, macroeconomic policy can be expected to improve, allowing for more realistic prices and better macroeconomic conditions (Krueger, 1974). Finally, the export sector can have a stimulating influence on the productivity of the economy as a whole by spillover effects on other sectors.[4]

9.2.2 The decision to export and how much to export

Turning to the theoretical determinants of becoming an exporter, and also of determining how much of output to export, there are two basic sets of explanatory variables: those related to sunk costs and those related to the future profitability of exporting. The decision to become an exporter has theoretically and empirically been shown to depend to a large extent on additional sunk costs of selling goods in the foreign market (e.g., Roberts and Tybout, 1997). Sunk costs may include expenses related to establishing a distribution channel and modification of commodities to foreign tastes. Trade liberalisation may increase the future profitability from exports by increasing the prices of tradable relative to non-tradable goods. However, the improvement must exceed the sunk cost to make a firm start to export, otherwise the firm will remain oriented towards the domestic market. Relatively more technically efficient plants may self-select into the export market because they expect the future returns are greater than the entry-costs.[5] The end result is that in a sample of non-exporting firms within the same industry, the more technically efficient firms should be more likely to become exporters.

Given that a firm is an exporter, the next question is how much of output to export. The set of explanatory variables related to future profitability may be equally relevant for the decision of how much to export. Of central concern here is the real exchange rate. A real depreciation is expected to increase the volume of exports as well as the export share. However, the sunk cost variable set will only be relevant if there are additional fixed costs related to increasing the export volume.

Both the export-led growth argument and the self-selection argument predict that exporting firms ought to be more technically efficient than non-exporters. In the influential Heckscher–Ohlin model of comparative advantage the focus is not on technical efficiency in production, but on the mapping from factor proportions to trade patterns. If the comparative advantage model is correct then firms producing commodities that use intensively the country's abundant factor should have a higher probability of being exporters compared with firms using the scarce factor intensively in production. If, for instance, a country is labour-abundant compared with its trade partners, the Heckscher–Ohlin model predicts that firms' probability of being exporters increases with the degree of labour-intensity in production.

9.3 Trade reform in Kenya

In this section, we discuss the trade reform programme implemented in Kenya and provide some indicators of the reform progress.

9.3.1 Kenyan trade reform: contents, implementation and impact[6]

The situation before the trade reform was that of inward orientation and excessive protection of the domestic industry. Quantitative restrictions on imports and high and variable tariffs protected local producers. There were also subsidised interest rates and tax exemptions to promote capital accumulation. Instead of fostering industrialisation, which was the aim of the import-substitution strategy, this policy led to scarcity of foreign exchange and poor growth. It provided protection for domestic inefficiency, encouraged overinvoicing, and led to the importing of more expensive goods. Most consumer goods imports were prohibited, while most intermediary inputs and capital were allowed unrestricted entry.

When exports to the East African Community (EAC) began to fall, some interest in the promotion of non-traditional exports emerged. The EAC was dissolved in 1977 and the regional market shrank. To compensate for the loss in income, a mechanism to reimburse eligible exporters for duties paid on imported inputs was introduced. This export compensation scheme suffered from problems, the main one being that the reimbursement did not reflect the duties and indirect taxes actually paid. There were also a number of obstacles facing exporters. The freedom to choose between domestic and imported inputs was limited; there was uncertainty about the cost of and access to imported inputs; and exporters had no priority in foreign exchange allocation.

The consequences for the manufacturing production were inefficiency, inward orientation, and high cost. The volume of manufacturing exports as a share of output fell, and so did imports. At the same time, domestic prices for manufacturing goods rose faster than import prices, leaving the manufacturing sector in an unfavourable competitive position.

However, in the 1980s there was a shift in trade policy towards openness. Sessional Paper No. 1 of 1986 (Republic of Kenya, 1986) outlined the importance of outward orientation to promote growth and economic development, and thus provided a turning point in terms of development philosophy. Starting in 1987, a second trade reform was gradually implemented.[7] The reform had two major components: export reform and import liberalisation. The former mainly comprised measures to improve support to the exporters and consisted of liberalising and streamlining the export licensing process. There was also increased direct support to exporters by establishing schemes for ensuring duty-free access to imported inputs and by providing subsidies for technical assistance. The import liberalisation started with a replacement of quotas by tariff-equivalents and was followed by a reduction of tariffs. Import licences, however, remained as a requirement, although initially these were easily obtainable.

Unfortunately, the export reform was not as substantial as was needed. Direct restrictions on exporters were greater than the direct support available to them. In fact, only two, poorly functioning, facilities were implemented: an export-compensation scheme and a manufacturing-under-bond facility. The idea of these facilities was to reimburse exporters for duties paid on imported inputs. The export compensation was often delayed and at that time few exporters actually benefited from it. Even less benefit was received from the manufacturing-under-bond facility, which was available only to firms specialising in export production.

The trade liberalisation led to a substantial decrease in nominal protection. However, effective rates of protection (ERP) between 1990 and 1992 were still quite high, although there was a tendency for reduction over time. Concentrating on the food, textile, wood and metal sectors, Table 9.1 shows that in 1992 the highest ERP was for garments (nearly 100 per cent), while the lowest ERP was for bakery products (31 per cent). For the other items, ERP ranged from 35 per cent to 58 per cent. Except for garments, there was a reduction in ERP for all sub-sectors between 1990 and 1992. On the negative side we note that the reduction of ERP in some cases has been minuscule and that overall the reduction was not impressive.

The trade reform as a whole could not be sustained because insufficient exchange rate adjustments undermined import liberalisation and inefficient fiscal adjustment impeded a substantial investment response. The problems were mainly concentrated on the import liberalisation part of the reform. While tariffs had been reduced, there was a reversal regarding the import licensing system. The reason was that import demand outpaced supply at prevailing exchange rates. Furthermore, the reduction of aid flows in the beginning of the 1990s reduced the availability of foreign exchange. This resulted in a lengthening of the queue for import licences and the applicants, apart from having to wait longer, also found that they were allowed to import a smaller amount than they had applied for. This was the way to balance actual imports and available foreign exchange. Thus the import regime once again became more restrictive.

Table 9.1 Effective rates of protection for chosen sub-sectors, 1990–92 (%)

Manufacturing production	*1990*	*1991*	*1992*
Food processing	51.0	50.6	44.1
Bakery products	42.9	40.7	30.9
Textile raw materials	41.1	41.1	35.9
Finished textiles	60.0	60.3	57.5
Garments	77.1	79.9	99.7
Leather and footwear	43.0	43.3	40.5
Wood products	48.2	47.3	43.4
Metal products	46.9	46.1	40.4

Source: UNDP/World Bank (1993).

9.3.2 Macro-evidence of manufacturing exports in Kenya

The importance for aggregate GDP growth from manufacturing exports growth is amply illustrated by the growth of East Asian countries, but also by, for instance, Mauritius. However, in Kenya, since independence in 1963, incentives have been strongly biased against exports, especially manufacturing exports. Consequently, Kenya has had difficulties in switching from primary exports to manufacturing exports. With the exception of horticulture, total exports have stagnated since the break-up of the EAC in 1977. The main recipients of Kenya's exports in 1995 were other African countries, primarily the PTA, which accounted for almost half of foreign absorption. Uganda and Tanzania are the largest African importers of Kenyan goods. The industrialised world absorbed slightly more than a third of Kenya's exports, and here the EU is the most important market. Within the EU, the most important recipient of Kenyan exports is the UK, followed by Germany and Italy. Hence, it remains true that south–south trade is the dominating type of trade in the Kenyan case (see Figure 9.1). Moreover, the range of products exported is still relatively narrow, compared with those of other low-income countries. As discussed above, two main reasons for this somewhat disappointing performance can be identified: the trade regime, and the macroeconomic and regulatory environment.

The picture changes when we turn to manufacturing exports alone. The market structure has remained unchanged since the 1970s, meaning that most manufacturing exports go to developing countries, with 90 per cent being directed to Africa and the Middle East. The share of manufacturing exports going to industrialised countries is small and highly variable.

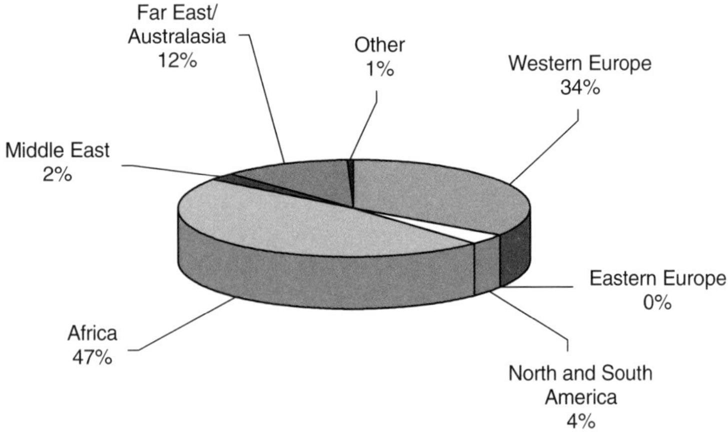

Figure 9.1 Direction of Kenyan exports, 1995
Source: *Statistical Abstract* (1996).

Lately, as a share of total exports there has been a remarkable increase of manufacturing exports over time. From having constituted approximately 17 per cent of total merchandise exports in 1975, its share amounted to almost 30 per cent in 1995 (see Figure 9.2). The major increase took place after the implementation of the second trade liberalisation in 1987. This is an indication that the trade reform efforts have started to pay off. We also see that the first attempt in 1980 had no significant effect on the manufacturing exports' share of total exports.

As a share of manufacturing output, manufacturing exports declined steadily from 1975 to the trade liberalisation in 1987 (see Figure 9.3). Thereafter, export increased its share of output. However, the export share is still very low and has yet to break the 10 per cent level. The composition of exports has changed lately from total domination by chemicals and cement to of other products having increasing significane, such as vehicles, printed matter, apparel and furniture.

We conclude this section with a graph on the development of the change of the real effective exchange rate and the change of manufacturing export share of output (see Figure 9.4). There is a fairly close correlation (although with a short lag) between the two variables with a real depreciation provoking an increase in the export share. A real appreciation, likewise, leads to a reduction in the share of output of manufacturing exports. It therefore seems to us that one possible tool for increasing the export share is that leading to real depreciation.

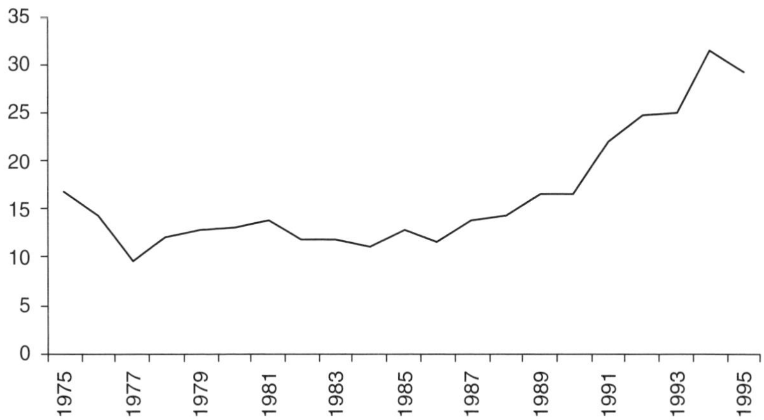

Figure 9.2 Manufacturing exports as a share of total exports, 1975–95

Source: *Statistical Abstract* (various issues).

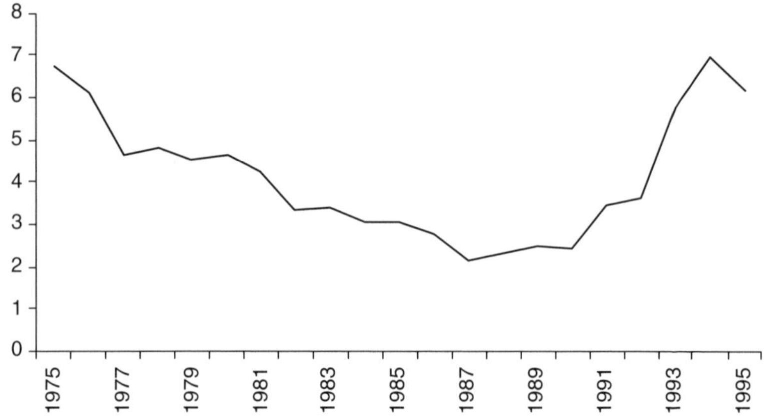

Figure 9.3 Manufacturing exports as a share of manufacturing output, 1975–95
Source: *Statistical Abstract* (various issues).

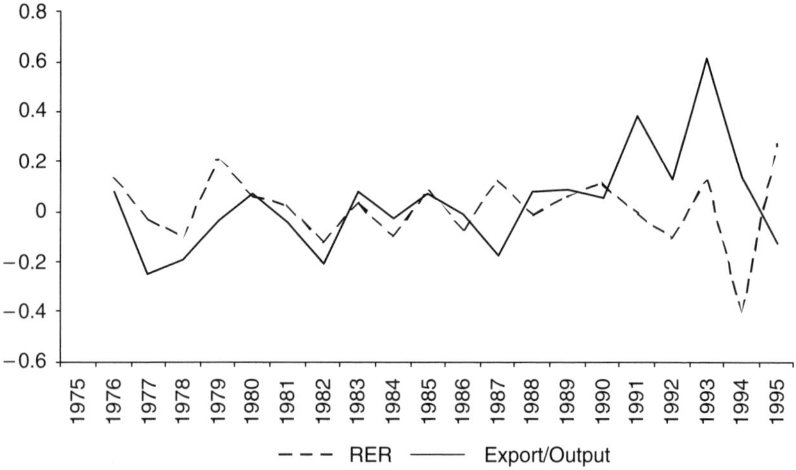

Figure 9.4 Changes in the real effective exchange rate and the manufacturing export Share, 1976–95

Source: Own calculations using data from IMF's *Direction of Trade Statistics* (various issues).

9.4 Micro-evidence in Kenyan manufacturing

In this section, we start with an analysis of the export performance for the four sub-sectors covered by the survey. To be able to draw inferences for the true population of the four sub-sectors, all calculations have been weighted to solve for the stratification bias in the data. Two measures of export performance are used: proportion of output exported and the number of firms exporting with respect to different firm characteristics. Thereafter, we briefly discuss entry and exit of firms into/out of the export market.[8]

9.4.1 Export performance

Table 9.2 shows the proportion of exporting firms as well as the proportion of output exported. Starting with the proportion of exporting firms, for the four sectors export participation increased dramatically between 1993 and 1995. The increase in export participation was shared by all industries, but was most significant for metals and textiles. This increase in export participation can most probably be explained by the trade reform.

For the four sectors taken together, the proportion of exporting firms increased, from 26 per cent in 1993 to about 36 per cent in 1995. Impressive increases were found for the textile sector (100 per cent) and the food and metal sectors (approximately 33 per cent), while the proportion of exporting firms in the wood sector increased only marginally.

Also, the proportion of exports increased from 24 per cent to 28 per cent between 1993 and 1995. The increase in export shares was highest in metals and textiles, while the increase in the wood sector was only marginal, and that of the food sector contracted. Hence the exporting firms in the food sector seem to have decreased their export orientation over time. Nevertheless, overall one may conclude that the Kenyan trade reform seems to have paid off both in terms of proportion of firms exporting and the export share.[9]

The observed increase in the share of output exported may originate both from new entrants into the export market and from increases in export shares from existing exporters. Table 9.3 reveals that the mean export share for exporting firms, for the whole sample, increased between the first and third wave from 32 per cent to 41 per cent. With this indicator

Table 9.2 Proportion of firms exporting and proportion exported by sector, 1993–95

Sector	*Proportion of exporting firms*			*Export share*		
	1993	*1994*	*1995*	*1993*	*1994*	*1995*
Food	0.346	0.352	0.444	0.336	0.355	0.255
Textiles	0.130	0.195	0.265	0.118	0.255	0.325
Wood	0.121	0.164	0.156	0.153	0.224	0.173
Metal	0.279	0.269	0.380	0.155	0.186	0.326
All	0.260	0.280	0.358	0.242	0.299	0.280

Note. Export share is calculated as the ratio between the sum of exported output and the sum of total output. Weights have been used for calculations. Formal firms only.

Table 9.3 Mean export share for exporting firms, 1993–95

Sector	*Proportion exported*		
	1993	*1994*	*1995*
Food	0.459	0.598	0.595
Textiles	0.193	0.220	0.343
Wood	0.253	0.165	0.230
Metal	0.188	0.228	0.331
All	0.319	0.370	0.409

Note: Weights have been used for calculations. Formal firms only.

of export performance, the metal and textile sectors almost doubled their mean export share. The food sector's export share increased by a quarter, while the wood sector was the only sector where exporters decreased the proportion of export sales during the sample period.

To get a picture of the relationship between exports and firm characteristics, Table 9.4 presents export performance by sector, location, firm size, and ownership structure. We see that, with respect to sectors, about half of the exporting firms in food, textiles, and wood direct some of their exports to non-African countries. Interestingly, almost all exporters from the metal sector direct their exports within Africa. This reflects the typical trade pattern for developing countries; manufacturing exports to developed countries consist of relatively labour-intensive goods, while trade between developing countries consist of relatively capital-intensive products.

Of the firms located in Nairobi, 36 per cent are exporters. For Nakuru the figure is slightly lower, while among firms in Mombasa 19 per cent are exporters. In Eldoret only 4.5 percent of the firms export. This may suggest that distance to ports as well as the capital city, where, for instance, government, judiciary and business centres are situated, affect the decision to export.

Comparisons of export participation between size groups reveal that the bulk of the exports in Kenyan manufacturing originates from large and medium-sized firms. As much as 66 per cent of large firms are exporters, while more than 95 per cent of small the firms sell solely to the domestic market. Ignoring the group of small firms due to the very few exporters in that group, it appears that the firm export share increases with size, while the propensity to export outside Africa is equal between these two size groups.

Finally, data on ownership provide interesting results. Only 20 per cent of the firms with exclusively domestic ownership export. The highest proportion of exporters is found in the category 'Mixed Ownership'. Domestically owned exporters, however, seem to export a larger part of output compared

Table 9.4 Export performance by sector, location, firm size, and ownership, average of 1993–95

Category	*N*	*Proportion of exporters*	*Proportion of exporters exporting outside Africa*	*Proportion of output exported for exporting firms*
Sector:				
Food	97	0.378	0.577	0.544
Textiles	113	0.197	0.557	0.279
Wood	120	0.147	0.450	0.209
Metal	113	0.309	0.005	0.256
Location:				
Nairobi	304	0.360	0.372	0.293
Mombasa	55	0.188	0.523	0.432
Nakuru	45	0.307	0.677	0.776
Eldoret	39	0.050	1.000	0.570
Size:				
Small	141	0.043	0.634	0.383
Medium	157	0.338	0.427	0.287
Large	145	0.658	0.418	0.400
Ownership:				
Foreign ownership	48	0.552	0.161	0.281
Mixed ownership	57	0.706	0.388	0.277
Domestic ownership	324	0.195	0.554	0.431
All	443	0.299	0.435	0.369

Note: N = Number of observations. Weights have been used for calculations. Formal firms only. For firm size, small refers to firms with less than 21 employees, medium-sized firms have from 22 to 75 employees, and large firms have more than 75 employees.

with exporting firms with some foreign ownership. Table 9.4 also reveals that a relatively large part of domestic-owned exporting firms tends to sell to markets outside Africa, while firms with foreign-owner interests to a higher degree tend to specialise their exports within Africa. A plausible explanation for this is that one common purpose of foreign direct investments in Kenya is to avoid trade barriers 'protecting' Kenya and other African countries belonging to regional trade areas (e.g., PTA and COMESA).

Several clear patterns emerge from the comparison of firm characteristics between exporters and non-exporters presented in Table 9.5. Technical efficiency, obtained from Data Envelopment Analysis (DEA), is higher for exporters in all sectors except for the textile sector.[10] However, the difference is statistically different only for the food and wood sectors. When all firms are taken together there is no statistically significant difference. With output as a proxy for size, exporting firms are almost six times larger than non-exporters. This difference in mean is statistically significant for all

Table 9.5 Characteristics of exporting firms in Kenya, 1993–95, and the ratio between the mean of exporting firms and non-exporting firms

Sector	*Technical efficiency*	*Output*	*Capital productivity*	*Labour productivity*	*Capital–labour ratio*
Food	1.58***	2.06**	0.89	1.08	0.87
Textiles	0.98	10.22***	0.30*	1.66**	4.09***
Wood	1.63***	7.93***	1.03	2.28***	0.95
Metal	1.19	14.79***	1.57	1.67**	1.20
All	1.19	5.73***	0.65*	1.91***	1.68***

Note: ***, **, and * indicate significant difference in mean at the 1 per cent, 5 per cent, and 10 per cent level, respectively. Weights have been used for calculations. Formal firms only.

Table 9.6 Firm transition rate in the export market between 1993 and 1995

1993 status	*1995 status*	*Food*	*Textiles*	*Wood*	*Metal*	*All*
No exports	No exports	0.846	0.905	0.889	0.824	0.872
	Exports	0.154	0.095	0.111	0.176	0.128
Exports	No exports	0.000	0.143	0.222	0.000	0.081
	Exports	1.000	0.857	0.778	1.000	0.919

Note: Formal firms only.

industries. Mean capital productivity is lower for exporters in textiles, the only sector in which the difference is statistically significant, but also for the sample as a whole. The low capital productivity in the textile sector among exporters may be explained by relatively capital-intensive technology. The mean of labour productivity is higher for the exporting firm group in all sectors and the difference is statistically significant in the textile, wood and metal sectors. The mean capital–labour ratio among exporters is more than four times the one for their domestically oriented counterparts in the textile sector, while the difference is not statistically significant in the other sectors. For the sample as a whole, however, exporters tend to have a higher capital–labour ratio.[11]

Finally, a few words on the entry and exit of firms into/out of the export market. Since firms that exited the sample between the waves were replaced, the increased export participation in the full sample may not appropriately reflect new entrants into the export market. Excluding firms that are not observed over all three years from the analysis provides a more reliable picture of entry and exit in the four sectors.[12] Table 9.6 shows the firm transition rate in the export market for a balanced sub-sample.

The top row indicates that of the firms that did not export in 1993, around 87 per cent of them remained non-exporters in 1995. The proportion of non-exporting firms that entered the export market was highest in the metal sector with almost 18 per cent, while less than 10 per cent of domestic-oriented firms in the textile sector switched to the export market. For the plants initially in the export market, the proportion of firms that remained in the export market varies from 78 per cent to 100 per cent across sectors. For the whole sample, the proportion of non-exporters that changed status to exporters was slightly higher than the proportion of exiting firms. With the caveat of examining a short sample period, it appears that once a firm has invested in entering the export market, it tends to stay there.

So far, we have provided only a partial analysis. Since the variables explaining the export decision and the export share may be correlated one must be cautious when interpreting these results. For instance, it is difficult to distinguish between the effect from technical efficiency and size on the decision to export due to the correlation between these two variables. With a multivariate analysis we are able to better isolate the effect from each of these variables. The next section presents the results from the econometric analysis.

9.5 The determinants of exports: an econometric analysis

Having provided descriptive statistics on exports, we now try to establish determinants as to why firms become exporters as well as to what factors determine the export share. We estimate two models; the first one explains the decision to export using probit regression, while the second one explains the share of output exported (estimated using OLS).

For both the decision to export and the export share-regressions we have two basic sets of explanatory variables. The first basic set reflects factors that are thought to determine the profitability of being an exporter. If the profitability of exporting increases with the export share, this set of variables is valid for both regressions. The second set of explanatory variables for the first regression reflects the additional fixed one-time costs involved in the process of entering the export market. For the second regression, however, the second set of variables is used to test whether there are additional sunk costs involved in increasing the export share.

Factors that determine firms' future profit from exporting include their size, the firms' technical efficiency, and their geographic location. Firm size, measured by the number of employees and the level of capital stock, may be important for several reasons. For instance, exporting may give the firm higher marketing costs than domestic sales, and the larger the firm the lower the average cost of exporting (Bigsten *et al.*, 1998). More related to sunk costs, firm size may also serve as a proxy for the magnitude of the firms' resources that are important for the decision to enter into the international markets. Examples are know-how, technology level, and access to credit.

Firms' technical efficiency may be positively related to the decision to export. It may be shown that relatively more productive plants self-select into the export market because the returns to doing so are relatively high for them (Clerides, Lach and Tybout, 1998). Geographic location, defined as a dummy variable taking the value of unity if the firm is located in either of the two largest cities (Nairobi or Mombasa) and zero otherwise, is thought to capture factors that influence transport costs and infrastructure. Some locations may be less appropriate to serve as a base for exporters. Location in the capital city may act as a proxy for infrastructure and business services, while location in Mombasa, since it is the main port in East Africa, lowers transport costs.

The presence of sunk costs provides an entry barrier into the export market. These might include expenses related to establishing a distribution channel, or production costs to modify domestic products to foreign tastes. These costs may vary with the skill of the staff, firm age, and ownership structure of the firm. Assuming that wages reflect an employee's productive ability, skill is measured by the wage cost per employee. The skill intensity of operations may capture the potential for technological activities, such as research and development. Furthermore, a high educational level within a firm facilitates international contacts. Firm age may capture the extent of a firm's learning experience. Age may also capture the cost difference among producers. If market forces select out technically inefficient producers, then older plants will tend to be more competitive in the world market (Roberts and Tybout, 1997). The structure of ownership may be important for the cost to access foreign markets. The importance of foreign ownership in the manufactured exports in many Least Developed Countries reflects the advantage of proprietary information, as well as special access to marketing networks abroad (Berry, 1992).

For both models, the suspected simultaneity bias caused by potentially endogenous explanatory variables, such as technical efficiency, labour, capital and average wage, is rectified by using the first lag of these variables. However, this is a costly procedure in terms of loss of degrees of freedom because we lose more than half the sample. Therefore, we also present regression results obtained without lagged regressors.

Industry dummies are included to capture unobserved industry-specific variables, such as product characteristics and the extent of domestic and foreign competition. The time effects are thought to pick up the influence of time-varying macro-variables, such as credit-market conditions, exchange rates, and trade policy conditions.

9.5.1 Estimation results

The results from the export decision and export share regressions are presented in Table 9.7. The columns labelled Model I give the parameters (marginal effects in the case of the export decision) for the specification

Table 9.7 Parameter estimates of export decision and export share models

Variables	*Export decision*		*Export share*	
	Model I	*Model II*	*Model I*	*Model II*
Constant	−0.843*	−1.090***	1.669***	2.222***
	(1.956)	(3.534)	(3.865)	(3.081)
Variables related to future profitability				
ln(Technical Efficiency)	0.114***	0.114	0.059	0.028
	(2.198)	(1.045)	(1.117)	(0.350)
ln(Employment)	0.212***	0.251***	0.085**	0.105*
	(5.258)	(3.221)	(2.151)	(1.807)
ln(Capital)	0.009	0.041	−0.062***	−0.062
	(0.378)	(0.816)	(2.645)	(1.482)
ln(Wage)	0.009	0.002	−0.020	−0.084*
	(0.263)	(0.026)	(0.661)	(2.017)
Variables related to sunk costs				
ln(Firm age)	−0.116**	−0.177*	−0.090*	−0.139**
	(2.285)	(1.891)	(1.932)	(2.024)
Big city	0.153*	0.121	−0.183*	−0.110
	(1.778)	(0.789)	(1.703)	(0.830)
Mixed ownership	0.318***	0.643***	−0.060	0.061
	(3.421)	(3.285)	(0.909)	(0.688)
Foreign ownership	0.232**	0.331**	0.018	0.096
	(2.559)	(2.176)	(0.227)	(0.933)
Control variables				
Food sector	−0.158*	−0.422***	0.252***	0.325***
	(1.776)	(2.217)	(3.136)	(2.657)
Wood sector	−0.318***	−0.480***	−0.128	−0.195
	(3.735)	(2.990)	(1.448)	(1.629)
Textile sector	−0.389***	−0.539***	−0.098	−0.171
	(4.203)	(2.766)	(1.040)	(1.298)
Wave II	0.110		−0.011	
	(1.577)		(0.170)	
Wave III	0.123*	0.028	0.008	0.024
	(1.793)	(0.283)	(0.119)	(0.335)
N	324	156	104	56

Note: ***, ** and * indicate statistical significance at the 1 per cent, 5 per cent, and 10 per cent levels, respectively. For export decision only the marginal effects are presented. For the export share equation, Heckman's two-stage estimation procedure produced an insignificant inverse mills-ratio, hence only the OLS results are presented.

with only contemporary explanatory variables, while the columns labelled Model II present the parameters (again the marginal effects for the export decision) for the specification with lagged independent variables.

With regard to the decision to export, of the explanatory variables related to future profitability of exporting, technical efficiency and firm size are statistically significant at the 5 per cent level in Model I. The effect from lagged technical efficiency is, however, insignificant at conventional levels.[13] This result indicates that firm size is an important determinant for the decision to export, while evidence of a positive effect from technical efficiency to the export decision is weaker. However, if firm size is driven by technical efficiency, as the theoretical model by Jovanovic (1982) suggests, the effect from technical efficiency may be underestimated.[14] The positive effect from employment, together with the insignificance of capital, suggests that the probability of exporting increases with labour-intensity. Since Kenya, compared with industrialised countries, is labour- abundant these results support the predictions of the Heckscher–Ohlin model. Location in a big city has a positive effect on the export decision. However, the parameter is only statistically significant in one of the models and only at the 10 per cent level.

Of the variables related to sunk costs, ownership (mixed and foreign) shows the expected sign on the propensity to export. Furthermore, the marginal effects are large, indicating that firms with mixed and foreign ownership have between a 23 and a 32 per cent higher probability of exporting than firms with only domestic owners (Model I). For Model II, the effect from mixed and foreign ownership is even larger (33 and 64 per cent higher probability, respectively). The unexpected effect from firm age may be explained by the possibility that relatively young firms utilise more recent technology, while older firms are stuck with relatively obsolete physical capital. Thus, in contrast to our expectations, older firms may be less competitive in the international market. Another explanation, related to the aforementioned one, is that old firms may be operating in the spirit of import-substitution policy: that is, they behave as if they were still overly protected. Therefore, they do not update their technology as rapidly as younger firms.

Turning to the proportion of output exported, in Model I we find a positive effect from labour and a negative one from capital, while in Model II the effect from capital is statistically insignificant. This may suggest that labour-intensity increases the export share, which is in line with what to expect from the theory of comparative advantage; Kenyan firms should, in general, export labour-intensive products. The total effect of a proportional change in labour and capital is, however, statistically insignificant, meaning that firm size is not relevant for the export share decision.[15] A negative statistically significant effect was obtained from firm age and skill. This result suggests that Kenyan exports are not very skill-intensive; instead, it is firms with goods produced using relative unskilled workers that export a large share of their output. Furthermore, it seems that relatively young firms are more outward-oriented than their relatively older counterparts. In none of the models has the technical efficiency a statistically significant effect.

Our overall conclusion concerning export decision and export share is that these two decisions have quite different determinants. For instance, ownership and firm size are important for the decision to become an exporter, but lack explanatory power for the decision as to how much of output to export. There is also some evidence that while more technically efficient firms tend to become exporters, their technical efficiency level does not affect their export share. A common determinant is firm age, which has a negative effect on both decisions. Finally, it is also worth noting that labour has a positive effect in all equations, thus supporting the notion of what the theory of comparative advantage dictates.

9.6 Conclusions

The purpose of this chapter has been to analyse the export performance in the Kenyan manufacturing sector. The study was based on descriptive and econometric analysis. We focused on micro-determinants of the decision to export and the share of output exported.

The descriptive analysis showed that, in general, exporting firms tend to be more technically efficient. Furthermore, they tend to operate on a larger scale then their domestically-oriented counterparts. Exporting firms are also more capital-intensive. Compared with domestically-owned firms, firms with some foreign ownership tend to a larger extent to direct their export to other African countries. There is also a larger proportion of exporters among these firms.

The results from the econometric analysis showed that the export-decision and export-share equation had different determinants. For the export-decision, whether owners were domestic, foreign or mixed, and firm size were important, while these variables lacked explanatory power for the export share. There was also some evidence that while more technically efficient firms tend to become exporters, their technical efficiency level does not seem to affect their export share. Our results thus support the existing theories of the importance of future profitability and sunk costs of entering into the export market.

Finally, we also found that an increase in labour has positive effects on both the export decision and on the export share, while capital has no effect on either decision. This result is in line with the Heckscher–Ohlin theory of comparative advantage because it is mainly labour-intensive firms with a low-skilled workforce that export, and these firms also export a higher share of their output. Thus, to compete on the world market Kenya should allocate resources towards labour-intensive industry production. In this way, the country should be able to reap the benefits from export-led growth.

Notes

1. When this chapter was written both authors were at the Department of Economics, Göteborg University, Sweden. Mats Granér is now at the Ministry of Industry, while Anders Isaksson is at UNIDO. They can be reached at

mats.graner@industry.ministry.se and aisaksson@unido.org, respectively. Please note that the views expressed herein are those of the authors and not necessarily those of UNIDO.

2. According to Easterly and Levine (1997), the closed-economy policies adopted by many African countries have been the single most growth-costly error. For a survey of trade liberalisation, openness and growth, see Edwards (1993).
3. Two exceptions are Bigsten *et al.* (1998) and Granér and Isaksson (2001). At the macro-level, however, the issue has been extensively analysed. Examples are Wood (1994), Wood and Berge (1997), and Owens and Wood (1997).
4. For a summary, see Bhagwati and Srinivasan (1978).
5. For a formal analysis, see Clerides, Lach and Tybout (1998).
6. This section draws from UNDP/World Bank (1993).
7. There had already been a first trial in 1980. Reinikka (1994) provides an overview of trade reforms in Kenya.
8. Since informal firms are not allowed to export unless they turn formal, they have been excluded from our sample.
9. Table 9.2 does not seem to square well with Figure 9.3, which showed much lower export shares. The explanation for these diverging results is that in Table 9.2 we only cover four sectors, while Figure 9.3 contains the whole of the manufacturing sector.
10. DEA is a linear programming method to calculate technical efficiency (see Charnes, Cooper and Rhodes, 1978). Technical efficiency was calculated with the technology restricted to constant returns to scale. We used gross output and three inputs (labour, capital, and material).
11. In the econometric analysis, the conclusion regarding higher capital–labour ratio for exporters are altered in that labour-intensity appears more important for exporting. The reason for the diverging results is that in a univariate analysis, such as that in Table 9.5, a high capital–labour ratio may proxy for firm size and therefore show up as statistically important. However, when controlling for firm size, the capital–labour ratio loses its statistical importance. That labour-intensity is associated with exporting seems more reasonable for a country such as Kenya, which is richly endowed with labour.
12. This analysis would have benefited from a longer sample because it would have provided much more variability in the data.
13. The effect from firm size is measured as a proportional change in capital and labour. Since the sum of the parameters of labour and capital is positively statistically significant, we conclude that firm size has a positive effect on the export decision (the chi-square for Model I is 57.73, and for Model II 27.31).
14. Technical efficiency was statistically significant at the 5 per cent level in both models when the regressions were estimated without capital and labour.
15. The chi-square values for Models I and II, respectively, are 0.58 and 1.01.

References

Berry, R. (1992), 'Firm (or Plant) Size in the Analysis of Trade and Development', in G.K. Helleiner (ed.), *Trade Policy, Industrialisation, and Development: New Perspectives*, Clarendon Press, Oxford.

Bhagwati, J. and Srinivasan, T. (1978), 'Trade Policy and Development', in R. Dornbusch and J. Frenkel (eds), *International Economic Policy: Theory and Evidence*, Johns Hopkins University Press, Baltimore, MS.

Bigsten, A., Collier, P., Dercon, S., Gauthier, B., Gunning, J., Habarurema, J., Isaksson, A., Oduro, A., Oostendorp, R., Patillo, C., Söderbom, M., Teal, F., and Zeufack, A. (1999), 'Exports of African Manufactures: Macro Policy and Firm Behaviour', *Journal of International Trade and Development* 8.

Charnes, A., Cooper, W. and Rhodes, E. (1978), 'Measuring the Efficiency of Decision-making Units', *European Journal of Operational Research*, 2.

Clerides, S., Lach, S. and Tybout, J. (1998), 'Is "Learning-by-Exporting" Important? Micro-Dynamic Evidence from Colombia, Mexico and Morocco', *Quarterly Journal of Economics* 113.

Easterly, W. and Levine, R. (1997), 'Africa's Growth Tragedy: A Retrospective, 1960–89', *Quarterly Journal of Economics* 112.

Edwards, S. (1993), 'Openness, Trade Liberalisation, and Growth in Developing Countries', *Journal of Economic Literature* 31.

Granér, M. and Isaksson, A. (2001), 'Understanding the Link between Exports and Productivity: New Evidence from Kenyan Manufacturing', mimeo, Göteborg University.

Helleiner, G. (1995), 'Manufacturing for Export in the Developing World: Problems and Possibilities: Introduction', in G. Helleiner (ed.), *Manufacturing for Export in the Developing World: Problems and Possibilities*, Routledge, London.

Jovanovic, B. (1982), 'Selection and the Evolution of Industries', *Econometrica* 50.

Krueger, A. (1974), 'The Political Economy of the Rent-Seeking Society', *American Economic Review* 64.

Owens, T. and Wood, A. (1997), 'Export Oriented Industrialization through Primary Processing?', *World Development* 25.

Reinikka, R. (1994), 'How to Identify Trade Liberalisation Episodes: An Empirical Study on Kenya', *Working Paper*, 94.10, Institute of Economics and Statistics, University of Oxford.

Republic of Kenya (1986), *Sessional Paper No. 1 on Economic Management for Renewed growth*, government printer, Nairobi.

Republic of Kenya (1996), *Statistical Abstract* 1996, government Printer, Nairobi.

Roberts, M. and Tybout, J. (1997), 'The Decision to Export in Colombia: An Empirical Model of Entry with Sunk Costs', *American Economic Review* 87.

UNDP/World Bank (1993), 'Kenya – The Challenge of Promoting Exports', *Report*, Trade Policy Division, Policy Research Department, Washington, DC.

Wood, A. (1994), *North-South Trade, Employment and Inequality: Changing Fortunes in a Skill-Driven World*, Clarendon Press, Oxford.

Wood, A. and Berge, K. (1997), 'Exporting Manufactures: Human Resources, Natural Resources, and Trade Policy', *Journal of Development Studies* 34.

10
Firm Growth

Renato Aguilar and Peter Kimuyu

10.1 Introduction

In this chapter we examine the growth of the manufacturing firms in Kenya to determine if there is a pattern in the growth process. Kenya's industrial base is narrow with a combination of a small group of large firms and an assortment of micro and small-scale enterprises (Economics Departments, 1994; Kimuyu, 1998). Part of the process of promoting industrialisation will include attempts not only to promote creation of new enterprises but also to help the existing ones to grow. Furthermore, increasing the pool of larger firms will boost employment in general by creating job opportunities more suitable for unskilled workers who are more abundant in countries such as Kenya (McCormick, 1993). The capacity of an industry to create employment depends on its skill requirements and labour intensity, both of which vary with enterprise size. Medium-sized industries, rather than the micro and small scale, are better placed to offer jobs for unskilled workers, since smaller firms may not have the resources train and therefore require workers with skills that can be applied directly. Encouraging enterprises to grow will pave the way for further job creation within the manufacturing sector and contribute to reduction of poverty.

There are other reasons why firm growth is a boon for an economy. Small enterprises rarely have either high capital productivity or the technical efficiency needed for industrial transformation. As studies referred to in McCormick (1993) and other results reported in Chapter 8 have shown, medium-sized enterprises perform better than micro and small-scale enterprises regarding productive use of capital and overall factor productivity. The greater efficiency of relatively larger enterprise is especially important for capital-scarce, labour-abundant developing economies such as Kenya. The growth of enterprises is also important from the point of view of technical improvement, since such improvement often occurs in the transition from very small to medium and large firms. Firms that remain small often lack capital for pursuing innovative ideas and continue to apply simple technologies, while larger enterprises tend to pick up more sophisticated productive processes. Small firms are not useful in technological progress. As argued in the literature, opportunities to exploit size, scale and growth economies should motivate enterprises to move into new products and expand production except where constant returns to scale dominate, in which case enterprises are indifferent to size.[1] Firms should therefore grow as markets expand either as a result of growth in local demand or penetration of foreign markets. In developing countries where the purchasing power is limited, enterprises are likely either to remain small or to diversify to other unrelated activities, dissipating limited managerial capability. Understanding the determination of firm growth is therefore important from a policy perspective.

The rest of the chapter is organised as follows. In section 10.2, we review the concept of firm growth and lay the groundwork for its estimation by paying attention to the structural laws that shape the direction of the impact of enterprise size and age on growth. The data and methods used form the subject matter of section 10.3. The estimation results are discussed in section 10.4, before concluding in section 10.5.

10.2 Operationalising the concept of firm growth

Analysing the growth of firms is almost intractable from both theoretical and empirical viewpoints. A major empirical problem is the specific measure of growth used as the object of analysis. The options that are mentioned in the literature include assets, profits, value added, employment, output and turnover (O'Farell and Hitchens, 1988; Ngugi and Ndeng'e, 1996). Nevertheless, there is growing literature on firm growth following a tradition that began in 1930 with the introduction of the 'Gibrat Law', whose main argument is that the initial size of a firm is immaterial with regard to the probability for a proportional increase in such size over a specified time interval. In other words, such a proportional increase, when it happens, is independent of the size of a firm. This law is the most widely used theoretical basis for empirical work on firm growth. Its alternative is the Jovanovic hypothesis which suggests the presence of an inverse relationship between firm growth and age or a disproportionate growth across firms depending on their age (Variyam and Kraybill, 1992). The Jovanovic hypothesis builds on an assumption about the existence of randomly distributed cost-function shocks with both transitory and time-invariant components. The latter represents managerial inefficiencies related to lack of experience in production. This cause of inefficiency is unknown to the enterprise at the beginning. Where production relations are such that output is a decreasing convex function of managerial inefficiency, the Jovanovic hypothesis leads to the conclusion that young enterprises are likely to grow faster than older ones.

Not all enterprises desire to grow. On the contrary, some are happy to remain small (O'Farell and Hitchens, 1988). A number of circumstances account for this outcome. For owner-managers, future enterprise goals are determined by family considerations and personal life styles. Other enterprises may prefer a strategy of survival rather than expansion, especially when economic independence is the primary preoccupation of the entrepreneur. Because a firm-growth strategy often makes dilution of ownership imperative, many owner-managers and partnerships may be unwilling to pay the necessary price for the growth of their enterprises. Enterprises may also shun growth for fear of attracting attention from labour unions and larger competitors, and proprietors may feel inadequate about meeting the management requirements generated by a growing enterprise such as those related to an expanded labour force, and organisational and supervisory complexities. Informal small enterprises may also shun growth for fear of attracting attention from tax authorities.

Even where entrepreneurs appreciate the benefits from growth, their response to risk and uncertainty dictates the eventual size of their enterprises (McCormick, 1993). Since entrepreneurs are generally risk averse, their response to risky situations has significant growth implications.[2] They may, for example, prefer meagre but certain returns to high but variable returns, or have a 'safety-first' response through which attempts are made to secure predetermined income levels. Part of this strategy necessitates the establishment of safety nets including product mixes, locations and enterprise sizes. Other strategies for addressing risk may include avoidance, spreading and compensation which translate into, among other things, minimising avoidable losses through adoption of strategies that inhibit expansion, such as conservative financial policies. In other words, an inherently risky environment is likely to condemn enterprises into remaining small.

O'Farell and Hitchens (1988) offer a number of theoretical perspectives on firm growth, namely, industrial economics, stochastic models, stage models and strategic models approaches. Under the industrial economics perspective, assuming declining short-run costs, savings in unit costs of production constitute the main motivation for firm growth. However, the rate of growth of enterprises is, under this theory, assumed to depend on both financial and demand factors, since capital is required to expand productive capacity and additional customers are needed to pick up increased production. Productive capacity is assumed to vary directly with the profit rate, while the profit rate is in turn assumed to vary with customer expansion and growth in the demand for the firm's output. The implication is that whereas there is a theoretical upper limit on the rate of expansion determined by these parameters, the absolute size of an enterprise is boundless.

The industrial economics theory of enterprise growth also embodies the 'Penrose effect', namely that the management sets limits to the rate of firm growth.[3] This is because, at any one point, there are limits to the extent to which an enterprise's management can increase its capacity. Hence lack of suitable management is an impediment to enterprise growth, so that even fast-growing enterprises are forced to slow down as they recruit new managers. Within this broad industrial economics perspective of firm growth, product diversification can also be viewed as a strategy for corporate growth.

The stochastic model of firm growth is based on casual observations that market concentration grows over time, and that the distribution of the sizes of firms is generally highly skewed with only a few large firms accompanied by a long tail of small ones. Among the skewed distributions that approximate these size distributions that are generated stochastically, the log-normal is the most familiar one (O'Farell and Hitchens, 1988). The literature also concludes that small enterprises may be perceived as passing through different stages, the most primitive of which stresses the personal attributes of the founder that are exemplified by an informal and simple management and organisational structure revolving around direct supervision and limited planning. The characteristics of the other stages are then propelled by attempts to address the management requirements of a more sophisticated organisation.[4]

Following the strategic model of growth, enterprises enter a new growth phase only after tackling impediments facing the previous phase. For small owner-managed enterprises, sustaining growth depends on the personal attributes of such owner-managers and the speed with which they can learn and adapt to the business environment.[5] From this perspective, the entrepreneur's personality has a definite impact on firm growth.

Firm growth has a spatial dimension in the sense that the impediments to growth vary in intensity and nature between different regions: for example, rural areas are unlikely to attract as much venture capital as the central business districts of major towns because lending institutions tend to be concentrated in the major trading centres. Furthermore, limited economic growth and prosperity in the peripheral areas reduces growth opportunities for small enterprises with a local focus. Firms located away from the main trading centres also face technological limitations, such as poor availability of trained manpower, inadequate opportunity for division of labour, greater need for vertical integration and denial of the benefits from specialisation. It is also suggested in the literature that a greater 'public–private services' gap and limited availability and access to assistance in the peripheral areas exert a negative effect on firm growth. The theoretical perspectives on firm growth are therefore many.

10.3 The data and models

The basic statistics of the data used in the analysis are presented in Tables 10.1 and 10.2. Table 10.1 shows the different sample sizes culminating in a panel with 340 observations. The main reference point in the data and analysis is the 1993 wave which has information on all the necessary variables. The average employment level was 8.9 employees and the average enterprise age 9 years. The managers were 39.1 years old, and exported a tiny 1.3 per cent of their output on average.

Table 10.1 Basic statistics in 1993

	N	*Mean*	*Standard*
Employment 1993	224	8.9	252.24
Age of the firm in years	224	9.0	28.98
% exported production	223	1.3	27.10
Age of the manager in years	191	39.1	12.20

Table 10.2 Some characteristics of the firms by ethnic origin of the manager

	Asian	*African*	*Other*	*Total*
Male manager	3.34	94.54	2.13	
	100.00	81.08	100.00	81.93
Female manager	0.00	100.00	0.00	
	0.00	18.92	0.00	18.07
Total	2.73	95.52	1.74	
Formal sector	65.04	32.24	2.72	
	100.00	1.26	10.34	3.77
Informal sector	0.00	99.08	0.92	
	0.00	98.74	89.66	96.23
Total	2.45	96.55	0.99	
Domestic	1.89	97.13	0.98	
	76.59	99.89	97.91	99.30
Foreign-owned	82.30	14.73	2.97	
	23.41	0.11	2.09	0.70
Total	2.45	96.55	0.99	
1–5	0.56	98.31	1.13	
	17.84	79.90	89.66	78.47
6–20	3.20	96.54	0.26	
	24.57	18.83	5.00	18.83
21–75	42.26	56.14	1.60	
	31.52	1.06	2.94	1.83
76–500	87.98	8.96	3.05	
	24.33	0.06	2.09	0.68
501+	22.85	75.49	1.66	
	1.74	0.15	0.31	0.19
Total	2.45	96.55	0.99	

Table 10.2 contains additional firm characteristics. There are no female managers among Asian firms. Asian firms were also only found in the formal sector, and they are highly represented among foreign-owned firms. While most small firms are either African-owned or managed, most of the large firms have Asian owners and managers.

It is not possible to choose a specification on a solid theoretical basis for a model explaining firm growth. Thus, the results are the outcome of a systematic empirical exploration of the data. Two sets of variables were related to growth. The first group includes the size and age of the firm. These are the traditional variables that are viewed as important in explaining firm growth. We included both variables and their squares in all the regressions. The squared terms help in exploring non-linearities in the firm growth models. The second group includes variables indicating the industrial branch, the ethnic origin of the general manager, the formality status and the ownership type (domestic or foreign owned).[6]

The data sets included information on the level of employment at different points of time. For example, there was information in employment not only for 1993 but also at the end of 1981 and 1986. Thus, it was possible to estimate rates of firm growth for 1981–86, 1981–93 and 1986–93. However, an examination of the response rates suggested that the best information for estimating annual growth rates was for the 1986–93 period. Thus the dependent variable in the regressions presented here is the logarithm of the yearly rate of growth obtained by comparing employment in 1993 with what it was in 1986.

For 1994 and 1995 estimations, the dependent variable was the logarithm of the yearly rate of growth since 1993.[7] This rate of growth was estimated by comparing employment data for 1993 with that for the other two consecutive years. The dependent variable used in the pooled estimations was the same as for 1994 and 1995. In any case, the estimated results draw from a variety of regressions obtained using ordinary least squares.

10.4 Empirical results

The results of the estimations for the basic model obtained over the entire sample, for which wave and for the panel are summarised in Table 10.3. Although the model does not fit the data very well, all four sets of estimates suggest a small positive effect following an inverted-U curve, both for firm size and firm age. The exception is the 1995 wave whose estimates elicit an opposite sign for firm age. In all cases, the statistical significance is low. In other words, the results suggest that, generally, firm growth is somewhat positively related to enterprise age and size, so that large older firm would continue to grow.

When some control variables were included in basic model, the overall fit of the model, as well as the significance of the estimated parameters,

Table 10.3 The basic model, full sample

Parameter	*1993*		*1994*		*1995*		*Panel*	
	Estimate	*t-index*	*Estimate*	*t-index*	*Estimate*	*t-index*	*Estimate*	*t-index*
Intercept	−0.072210	−2.093	0.189439	2.141	0.042667	0.420	0.123951	1.817
Employment	0.000283	1.565	0.001135	0.889	0.002584	1.581	0.001729	1.676
Employment squared	−0.000000128	−1.289	−0.000000493	−0.672	−0.000001239	−1.340	−0.000000788	−1.337
Age	0.004845	1.547	0.004812	0.421	−0.013063	−1.048	−0.005716	−0.667
Age squared	−0.000068896	−1.074	−0.000368	−1.257	0.000120	0.415	−0.00008338	−0.401
n	158	–	195	–	169	–	364	–
F, R^2	1.570	0.0394	1.107	0.0228	1.161	0.0275	1.898	0.0207
Prob > F	0.1850	–	0.3546	–	0.3300	–	0.1103	–

improved slightly (Table 10.4). The estimated parameters for 1995 and the panel are particularly significant. However, the second-degree parameters continue to show limited significance in all cases. The parameters corresponding to the effect of belonging to the textile and metal sectors, rather than being in the wood sector, are relatively large and significant. In some cases the effect of belonging to the metal industry as compared to the wood industry is also estimated with a satisfactory level of significance. However, the growth performance of enterprises in these sectors relative to those in the wood sector switches from negative to positive, so that firms in these sectors grew at a significantly slower rate than the wood-working firms in the 1986–93 period, but at a faster rate in subsequent periods.

More generally, some evidence supporting the existence of a statistical structure relating firm growth to firm size or firm age can be found. However, such evidence is rather weak. Furthermore, there is no clear evidence for associating firm growth with ethnicity, formality status and ownership. However, there is some evidence of possible differences in the growth performance of firms in the different sectors.

Additional analysis on the wood sector picks a positive effect of firm age on growth for 1993. This effect follows the usual inverted-U curve pattern. For the same sector, negative effects were estimated for 1995 and the panel. For the firms in the food sector, the model fits the data reasonably well only for 1994. The estimates suggest a positive non-linear effect of firm size on growth. However, only for the full panel were the estimates for this sector statistically significant. For the textile sector, the model does not fit the data for any sub-sample. This was probably on account of the fact that textiles were most affected by liberalisation of the Kenyan economy. For the metal sector, the model fits the data reasonably well for 1994 and for the panel.

A somewhat significant positive effect of firm age on the growth of metal-working firms is identified. This effect is particularly strong for the 1994 and 1995 waves, as well as for the pooled panel of the sector. Thus, the breakdown by branch suggests differentiated patterns of growth by branch with metal and wood showing an inverted-U relationship between growth and firm age. In the case of the food sector there is weaker empirical evidence in support of the relationship between firm growth and firm size. For the textile sector, no pattern could be identified.

An attempt to explore the impact of ethnicity on firm growth yielded very weak results, with the models poorly fitting the data for both Asian and African firms (Tables 10.5 and 10.6). Further analysis on the impact of formality status on firm growth reveals that the model fits the data of the sub-sample of formal firms better than the smaller sub-sample of informal firms (Tables 10.7 and 10.8). In the case of the formal sector there seems to be a positive effect of firm size. This effect was particularly significant in 1993 and 1994. However, there is no evidence of non-linearity. The results for the informal sector were rather poor, perhaps on account of the small

Table 10.4 Basic models with controls, the whole sample

Parameter	*1993*		*1994*		*1995*		*Panel*	
	Estimate	*t-index*	*Estimate*	*t-index*	*Estimate*	*t-index*	*Estimate*	*t-index*
Intercept	−0.118751	−1.221	−0.498013	−0.752	−0.486359	−0.701	−0.488172	−1.018
Employment	0.000409	1.572	0.004511	1.710	0.005691	2.018	0.004995	2.666
Employment squared	−0.000000229	−1.330	−0.000002965	−1.408	−0.000003563	−1.834	−0.000003194	−2.272
Age	0.003979	1.294	0.007775	0.647	−0.013813	−1.010	−0.005456	−0.602
Age squared	−0.000055873	−0.864	−0.000287	−0.918	0.000198	0.598	0.000005778	0.025
Food	−0.059769	−0.823	0.317156	1.904	0.063154	0.471	0.114454	1.113
Textiles	−0.131433	−5.842	0.260787	2.433	0.280842	2.424	0.309763	3.972
Metal	−0.102802	−3.924	0.180737	1.522	0.195848	1.686	0.199229	2.390
Asian	−0.003160	−0.054	0.048314	0.125	−0.239972	−0.526	−0.058999	−0.199
Informal	0.014621	0.274	0.596302	1.865	−0.018370	−0.047	0.292079	1.171
Domestic	0.138361	1.904	−0.139889	−0.247	0.392430	0.682	0.118845	0.293
N	144	–	183	–	157	–	340	–
F, R^2	4.631	0.2583	1.601	0.0852	1.327	0.0833	2.847	0.0796
Prob > F	0.0001	–	0.1097	–	0.2210	–	0.0021	–

Table 10.5 Ethnic breakdown, Asians

Parameter	*1993*		*1994*		*1995*		*Panel*	
	Estimate	*t-index*	*Estimate*	*t-index*	*Estimate*	*t-index*	*Estimate*	*t-index*
Intercept	−0.106120	−1.062	0.315764	0.648	−0.657946	−3.772	−0.232239	−0.889
Employment	0.001053	3.546	0.004659	2.330	0.001136	1.555	0.002877	2.771
Employment squared	0.000000503	−3.161	−0.000002952	−1.939	−0.000000575	1.187	−0.000001629	−2.185
Age	0.000716	0.105	−0.050041	−1.587	0.029911	2.663	−0.006546	−0.387
Age squared	0.000006504	0.066	0.000596	1.307	−0.000346	−2.155	0.000106	0.436
N	92	–	93	–	78	–	171	–
F, R^2	3.239	0.1295	3.599	0.1406	2.767	0.1317	3.158	0.0707
Prob > F	0.0158	–	0.0091	–	0.0336	–	0.0156	–

Table 10.6 Ethnic breakdown, Africans

Parameter	*1993*		*1994*		*1995*		*Panel*	
	Estimate	*t-index*	*Estimate*	*t-index*	*Estimate*	*t-index*	*Estimate*	*t-index*
Intercept	−0.029512	−0.368	0.198142	0.992	−0.235488	−1.000	−0.012560	−0.081
Employment	0.003069	2.223	0.019852	1.546	0.041492	3.691	0.031865	3.735
Employment squared	−0.000003577	−2.175	−0.000061597	−1.509	−0.000098295	−3.161	−0.000083577	−3.328
Age	−0.005142	−0.494	−0.016838	−0.444	0.006743	0.159	−0.005866	−0.205
Age squared	0.000300	0.944	0.000563	0.396	−0.000672	−0.459	−0.000084385	−0.082
N	46	–	81	–	72	–	153	–
F, R^2	1.801	0.1494	0.651	0.0331	4.113	0.1972	3.840	0.0940
Prob > F	0.1472	–	0.6279	–	0.0049	–	0.0053	–

Table 10.7 Formality breakdown, formal sector

Parameter	*1993*		*1994*		*1995*		*Panel*	
	Estimate	*t-index*	*Estimate*	*t-index*	*Estimate*	*t-index*	*Estimate*	*t-index*
Intercept	0.033049	0.511	0.189697	0.643	−0.402865	−2.898	−0.118004	−0.691
Employment	0.000265	2.448	0.001757	2.501	0.000749	1.822	0.001334	2.978
Employment squared	0.000046969	0.672	−0.000000697	−1.800	−0.000000304	−1.367	−0.000000534	−2.177
Age	−0.003933	−0.838	−0.025144	−1.185	0.019831	2.063	−0.002279	−0.189
Age squared	0.000046969	0.672	0.000121	0.384	−0.000240	−1.712	−0.000042621	−0.240
N	139	–	142	–	123	–	265	–
F, R^2	1.790	0.0507	4.350	0.1127	2.176	0.0687	3.660	0.0533
Prob > F	0.1345	–	0.0024	–	0.0758	–	0.0064	–

Table 10.8 Formality breakdown, informal sector

Parameter	*1993*		*1994*		*1995*		*Panel*	
	Estimate	*t-index*	*Estimate*	*t-index*	*Estimate*	*t-index*	*Estimate*	*t-index*
Intercept	−0.123972	−0.795	−0.084097	−0.300	−0.260598	−0.866	−0.118656	−0.566
Employment	0.028761	2.103	0.121290	2.097	0.060717	1.189	0.067959	1.790
Employment squared	−0.000931	−1.334	−0.005496	−1.724	0.000023425	0.010	−0.001163	−0.612
Age	−0.002290	−0.111	−0.019400	−0.421	−0.004866	−0.094	−0.009139	−0.260
Age squared	0.000216	0.321	0.000878	0.505	−0.000275	−0.153	0.000051940	0.041
N	19	–	53	–	46	–	99	–
F, R^2	2.450	0.4118	1.338	0.1003	4.069	0.2842	3.908	0.1426
Prob > F	0.0946	–	0.2696	–	0.0072	–	0.0056	–

informal sub-sample. However, the coefficients for the squared terms suggest that non-linearity cannot be ruled out for the informal manufacturing.

Analysing the impact of ownership on firm growth reveals that the models fit the smaller sub-sample of foreign firms better than that for domestic firms. For the latter, all the estimated parameters are insignificant, so that it is not possible to identify a statistical structure for this sub-sample. For foreign firms, however, a significant positive effect of firm size on firm growth was consistently elicited. Although the estimated parameters for the squared term were quite small, they were frequently statistically significant, confirming that the relationship between firm growth, age and size is not linear. This clear contrast between the growth behaviour of domestic and foreign-owned firms probably results from the greater ability of the foreign-owned firms to exploit economies of scale and technological changes. Better access to finance and technology would give foreign firms the ability to keep adjusting towards point of best practice and grow in the process.

The model also fits relatively well all sub-samples defined by the firm size breakdown. In the case of the micro enterprises with only 1–5 employees, the inverted-U shaped positive effect for firm size was evident. Moreover, the estimated parameters were relatively large. However, firm age does not seem important in explaining the growth of these micro enterprises. This might reflect the short life expectancy of the very small enterprises in Kenya. The general pattern for the micro enterprises is observable for the other sizes of sub-samples except that the estimated coefficients are smaller. In other words, the statistical structure for the micro enterprises is more revealing.

10.5 Summary

There does not seem to be evidence for the existence of some *general* relationship explaining firm growth in Kenya's manufacturing industry. However, such relationships exist for specific sub-samples and are particularly evident along the firm size and ownership strata. The relationship is weak for formality status and ethnicity. Generally, the results suggest that firm size has a positive role in explaining firm growth. Most often the effect has the shape of an inverted-U curve, suggesting that the growth path is not linear. In any case, the effect is small. Although these results contrast with earlier ones, they agree with more recent findings that identify some relationship between firm size and growth for specific segments of manufacturing.

The existence of *partial* relationships explaining firm growth underscore the idea that different groups and sectors are related in a differentiated way to the business cycle, technology, financing and the general environment determining productivity and performance of Kenyan manufacturing.

The confirmation of positive firm size–growth relationships, albeit non-linear, suggests the existence of a size threshold within which enterprise growth is maximised and the employment benefits of such growth are fully realised. Since Kenya's private sector is expected to generate the largest share of employment opportunities in the future, public policy will need to focus on creating incentives not only for encouraging the establishment of new enterprises, but also for pushing the existing ones toward the size threshold in which self-propulsion of growth dynamics is most likely to come into play. Part of that strategy has to include the reduction of the commercial uncertainties that reduce the survival rate and life expectancy of micro and small enterprises which dominate Kenyan manufacturing, in order to permit the growth benefits of enterprise longevity to accrue.

Notes

1. As argued in McCormick (1993), firms in the same sub-sector may use different technologies, so that it is not easy to identify optimal enterprise size from a purely technological viewpoint.
2. McCormick (1993) defines risk aversion as preference for riskless or low risk situations.
3. The 'Penrose effect', reported in O'Farell and Hitchens (1988), is attributed to Penrose (1959).
4. The assumptions about the actual number of stages and their specific attributes varies considerably, with each stage being either a cause or effect of another. See also O'Farell and Hitchens (1988).
5. The specific personal attributes mentioned in the literature include age, management style, decision making, education, training, values and attitudes. There are also external aspects to the business environment, such as suppliers, buyers, degree of competition, interest rates, fiscal policies such as those affecting corporate taxes, sectoral trends, exchange rate regimes and the prevailing political and legal conditions.
6. The observations on other categories were too few to permit meaningful exploration. The basic model was also estimated for different firm sizes.
7. The dependent variables were the logarithm of the yearly rates of growth plus one. The one was included to make the computation of the logarithm of zero unlikely.

References

Economics Departments (1994), 'Limitations and Rewards in Kenya's Manufacturing Sector: A Study of Enterprise Development', Department of Economics, University of Göteborg, Sweden and Economics Department, University of Nairobi, Kenya.

Kimuyu, P. K. (1998), 'Industrial Policies for the Twenty First Century: Productivity, Competitiveness and Export Participation', Paper presented during the Institute of Policy Analysis and Research National Seminar on 'Strategic Development Paths for Kenya in the 21st Century', organised by the Institute of Policy Analysis and Research, 15–16 April 1998, Safari Park Hotel, Nairobi, Kenya.

McCormick, D. (1993), 'Risk and Growth: the Dilemma of Nairobi's Small-Scale Manufacturers'. *Discussion Paper* No. 291, Institute of Development Studies, University of Nairobi.

Ngugi, J. K. and Ndeng'e, G. (1996), 'Determinants of Small Enterprise Growth in Kenya', Research Proposal submitted to the African Economic Research Consortium, Nairobi.

O'Farell, P. N. and Hitchens, D.M.W.N. (1988), 'Alternative Theories of Small Firm Growth: A Critical Review', *Environment and Planning* 20.
Penrose, E. (1959), *The Theory of the Growth of the Firm*, Oxford, Blackwell Scientific Publications.
Varyam, J.N. and Kraybill, D.S. (1992), 'Empirical Evidence of the Determinants of Firm Growth', *Economics Letters* 38.

Part V
Impact of Policy and Institutional Reform

11
The Impact of Structural Adjustment on Manufacturing

Jörgen Levin and Njuguna Ndung'u

11.1 Introduction

In 1980 Kenya became one of the first countries in Africa to sign a structural adjustment agreement with the World Bank. Kenyan policy-makers and the Bank saw this as an important step towards greater economic liberalisation, in an attempt to revive growth. However, in the first part of the 1980s, notably 1980–84, the government showed total lack of compliance, partly due to problems of design and timing and partly due to lack of commitment (Swamy, 1994). While the period 1985–91 showed some improvement in the main aggregates, macroeconomic stability remained difficult to establish. In a bid to speed up the economic reform, as well as the process of democratisation, donors withdrew quick-disbursing aid to Kenya in November 1991. The run-up to multiparty elections, held in December 1992, saw a great deal of money creation. Like most governments, Kenya wished to ensure that the election was not lost for lack of financial generosity on its part. In the period 1993–95, it embarked on short-term crisis management to deal with the monetary overhang from the election and the destabilising effects of the coffee boom. Drastic measures were also taken in liberalising both the current and capital account, which induced huge capital inflows. The impact on firms' operations became severe in several areas. The major problem during the survey period was access to credit and lack of demand following tight monetary policy. Increased competition became a severe problem, which was further aggravated by rising utility prices and deteriorating infrastructure. Discussions of infrastructure issues are presented in Chapter 12. The purpose of this chapter is to discuss the impact of the structural adjustment programmes on manufacturing performance. The chapter proceeds as follows: section 11.2 describes policy outcomes at the macro level during 1986–95. The third part highlights some of the findings from the survey. The chapter is concluded with a discussion of the major findings.

11.2 Macroeconomic adjustment and trade liberalisation, 1986–95

In the mid-1980s Kenya experienced a 'mini' coffee-boom and at the same time as it was trying to initiate long-term growth policies. These were outlined in Sessional Paper No. 1 of 1986, entitled 'On Economic Management for Renewed Growth'. The intention was to move away from a restrictive trade regime towards a more flexible system. Tariffs and exchange rate policies were important in the new system. A three-stage approach was adopted: first, quotas would be replaced by their tariff equivalents. To reduce costs for domestic industry, duties on industrial inputs would be lowered. In the third and final stage, duties on final products would be lowered, thus exposing domestic industry to competition from imports. Though implementation of the programme was initially delayed, some progress was made as foreign exchange allocation was improved, licensing for capital and intermediate goods relaxed, and the tariff system reformed (Reinikka, 1993).

However, the coffee boom was short-lived and, together with less favourable weather conditions and a major crisis in the banking sector, the economy was distressed once again. To restore fiscal and monetary discipline the government adopted a three-year stabilisation and structural adjustment programme, at the beginning of 1988. GDP continued to grow at around 5 per cent, and the internal and external balance improved slightly compared to the previous year (Table 2.4 in Chapter 2). Money supply increased by 6 per cent compared to 1987 (Table 11.1). Domestic credit to the government was reduced by 8 per cent while that to the private sector increased by 20 per cent. To broaden the adjustment process, a three-year Enhanced Structural Adjustment Programme replaced the earlier structural adjustment programme in 1989. The year saw continued implementation of trade reforms. Quotas were replaced by tariff equivalents, which raised the average tariff level. However, since the macroeconomic objectives were not fully met, inflation and both internal and external balances deteriorated further. Aid inflows enabled the country to avoid serious adjustment for the time being. At the end of the decade the resource deficit was 7.1 per cent, a result of rapid growth in private spending. Thus, although the adjustment programme succeeded in limiting government expansion, it failed to control private consumption and investment. Increased aid, coupled with trade liberalisation and a relaxed credit policy (as depicted by a declining cash-ratio in Table 11.1) towards the private sector, became balance of payments incompatible. To reduce pressure on the external sector the exchange rate needed to be depreciated further.

The movements of relative prices during 1986–89 are illustrated in Figure 11.1 and Table A11.1. In Figure 11.1 P_x/P_m and P_n/P_m are traced on the vertical and horizontal axes, respectively. The former are determined by the

Table 11.1 Exchange rates, inflation and other monetary indicators

	1986	*1987*	*1988*	*1989*	*1990*	*1991*	*1992*	*1993*	*1994*	*1995*
Real exchange rate*	−1.8	5.7	7.0	16.5	5.6	−1.5	−14.1	32.6	−5.6	62.3
Nominal exchange rate*	−1.5	2.9	12.6	16.1	11.5	16.6	29.0	88.2	−34.2	24.8
Consumer price index*	3.9	7.7	11.2	12.9	15.6	19.8	29.5	45.8	29.0	0.8
Domestic credit*	28.6	20.4	6.6	7.0	30.6	19.7	15.3	0.4	46.0	37.9
Domestic credit, government*	47.7	29.5	−8.0	−5.5	64.0	12.2	−0.8	−4.6	75.9	20.0
Domestic credit, private sector*	16.8	13.2	19.6	15.6	11.8	25.9	27.1	3.2	30.2	50.7
Money supply (M1)*	32.6	9.7	5.9	8.6	12.1	16.4	43.9	22.6	11.1	9.0
Money supply (M2)*	32.5	11.2	7.9	12.9	27.2	20.9	33.6	25.7	30.4	18.6
Money supply (M3)*	26.5	10.7	10.9	16.1	28.3	17.3	26.9	21.6	25.4	11.3
Discount rate for Treasury Bills	12.1	13.0	13.5	14.0	15.9	16.8	17.0	39.3	17.9	20.9
Cash-ratio	11.0	10.4	9.2	9.1	10.6	10.6	11.0	19.0	22.8	21.2

* Annual changes.
Sources: Republic of Kenya, *Economic Survey*, various issues.

international terms of trade and the trade restrictions imposed, whilst the latter are determined by the exchange rate, trade restrictions and domestic money supply (which determines the price level of non-tradables).[1] In Figure 11.1, 1986 is a point where both money and non-tradables markets are assumed to be in equilibrium. Assuming that the non-tradables market clears instantaneously, the *NN*-locus is drifting to the right during 1986–89, with P_n/P_m increasing. Expansionary fiscal and monetary policies, coupled with a less restrictive trade regime, induced private agents to reduce savings and increase private expenditures. There was a boom in building and construction during this time, which drove up the prices of non-tradables.

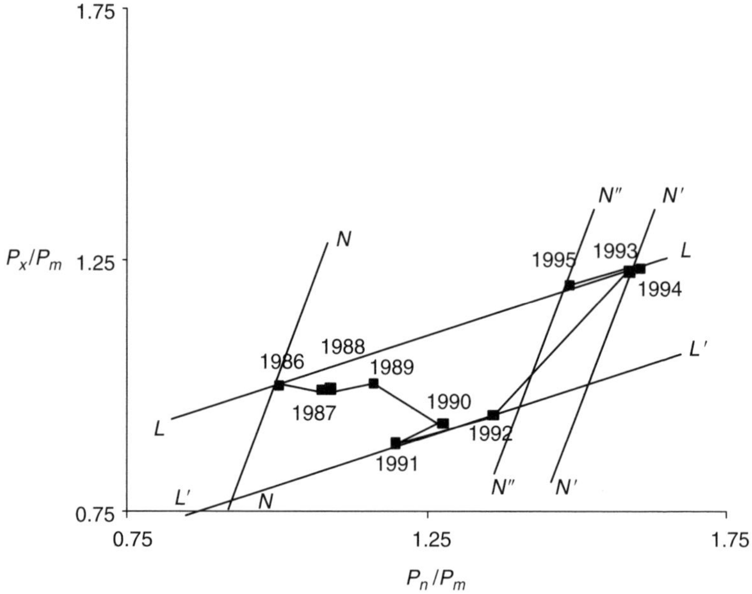

Figure 11.1 Changes in relative prices, 1986–95

In the money market expansionary fiscal and monetary policy shifted the *LL* curve downward to *L′L′* and with disequilibrium prices the money market moved into excess supply, implying a balance of payments deficit. Thus, the exchange rate was not devalued sufficiently in light of the external shocks and liberalisation measures of the period.

In 1990, terms-of-trade deteriorated further, growth was reduced to 4.2 per cent and current account improved slightly, but the deficit was still 6 per cent of GDP. The fiscal deficit also deteriorated, financed by heavy government borrowing from the banking system (see Table 11.1). This implied a large expansion in domestic credit. But while credit to the private sector and non-administrative public institutions increased by 11.8 per cent, public sector borrowing increased by 64 per cent. This, combined with changes in international reserves, led to a 27 per cent growth in money supply (M2), much higher than in 1988 and 1989. The depreciation of the shilling by 11.5 per cent helped offset the expansionary policies to some extent, but without entirely eradicating the effects of the excess money supply. The *NN*-curve continued to shift to the right. In contrast to the previously discussed period this was a result of increased government expenditures. The movement of relative price configuration to the south-east meant incentives for resources to move from exports and import substitutes to non-tradables.

During 1991, overall performance continued to be disappointing with three out of four performance criteria unmet: ceilings on net domestic assets of the banking system as well as net bank credit to the government were exceeded. The net official international reserves target was not met either, while the public sector deficit remained above 11 per cent of GDP. GDP growth declined to 2.1 per cent. In the relative price space shown in Figure 11.1, the *NN*-curve moved south-west between 1990 and 1991, the result of a general slowdown in economic activity, but with added effects emanating from increased tariff protection. However, the bulk of the liberalisation measures had been completed about this time. Quotas were replaced by tariffs and the latter rose on average, reaching their highest level during 1990–91. This raised prices for importables, P_m, and moved the economy closer to equilibrium in the money market. However, the inability to implement the programme efficiently implied that movements in relative prices were the opposite of what was expected. Thus, incentives for an export-led strategy were further eroded.

Following the government's increasing difficulty in implementing the adjustment programme, an unfavourable human-rights record and the wish on the part of donors to use aid leverage to push the issue of multiparty elections, Kenya's main donors withdrew quick-disbursing aid during 1991. ForexCs, which had been introduced earlier, helped to ensure a reasonable inflow of foreign exchange and became the major source of it for the private sector. This inadvertently became the first major move in currency market liberalisation and was also an important instrument in attracting capital from abroad. The introduction of ForexCs implied, however, that Kenya was operating under a dual exchange rate regime.

In early 1992 discussions resumed between the Government and the IMF, and a 'shadow' programme was agreed on. Its main ingredient was the establishment of a stable macroeconomic environment, but without additional funding from the donors. Thus, tight fiscal and monetary policies were to be its main feature. The economy was pushed into recession, growing at only 0.5 per cent in 1992. Besides the events of 1991, other contributing factors included bad weather, political uncertainty, ethnic clashes and policy mismanagement. In 1992 broad money grew by 34 per cent, as opposed to a target of 9 per cent. Real interest rates became more negative as inflation rose to 29 per cent. There was no resumption of aid.

In the relative price space, there was a move to the north-east (Figure 11.1). The *NN*-locus shifts to the right, reflecting slow growth of supply because of low investment levels in previous years but also due to the expansionary policies: real money balances increased. During 1992 there was also a reduction in the average tariff level: this might have had the effect of shifting the *NN*-curve further to the north-east.

By 1993, it was realised that drastic steps were needed to put the economy back on track. Tight fiscal and monetary policies had proved impossible to implement. The government announced that all private sector foreign exchange transactions would be via the 'second window', bought at a higher rate than the official one. Also, parts of the earlier licensing system were reintroduced. Although the interbank market was supposed to operate on a market basis, foreign exchange rationing remained. In response to the policy uncertainty characterised by high inflation, domestic agents shifted into dollar-denominated assets. The Central Bank's foreign exchange reserves were almost depleted as a result, with less than two weeks of imports. The spread between the official exchange rate and the interbank market was about 45 per cent, far above the recommended level. The Central Bank covered its losses through money creation. In a move to narrow the gap between the interbank and the official rate, the shilling was devalued twice altogether by a total of 56 per cent. The licensing system was abolished and government controls on imports were largely dismantled. The tariff structure was harmonised and tariff dispersion lowered (see Chapter 2).[2]

In October the dual exchange rate system was removed and the shilling began to float. These moves restored confidence reversing the flow of foreign exchange out of the country, and the currency started to appreciate steadily. A high yield on domestic Treasury Bills also attracted further inflows. This was used extensively to mop up excess liquidity along with a higher cash ratio, which reached 19 per cent in December 1993. Although the macroeconomic situation had improved during the latter part of the year, the overall performance remained mixed. GDP growth was about 0.3 per cent and investment continued to contract. A major factor was that performance in the agriculture sector was still unsatisfactory mainly due to the prolonged drought. However, improved terms-of-trade, due to the sharp increase in coffee prices, and huge inflows triggered by the high yield on Treasury Bills, led to a dramatic increase in foreign exchange reserves.

In 1994 the economy showed signs of movement towards internal and external balance. GDP grew by 3.0 per cent and investment showed positive growth for the first time since 1990. The return of good rains meant that agricultural output rose by 2.8 per cent. The tight monetary policy initiated in the second half of 1993 continued during 1994. Treasury Bills, although their rates have fallen by over 20 percentage points from 1993, continued to be sold, keeping interest rates high, while further increases in the cash ratio were made during the year. In December 1994 it had reached 22.8 per cent. While the year also saw stabilisation of the exchange and interest rates, the fiscal deficit remained at a high level.

The agricultural sector continued to perform well during 1995. There was also considerable improvement in manufacturing, its growth rebounding back to the 1991 level, but still below its performance in the 1980s. Altogether GDP increased by 4.9 per cent. Overall investment grew by about 17 per cent in 1995 (Table 2.5 in Chapter 2) where the biggest increase was found in the agriculture and manufacturing. In the latter, investment increased by 41 per cent. The public sector deficit/GDP ratio was reduced to about 2.1 per cent. A boost in revenue was the major contributing factor, a result of proceeds from privatisation and the general recovery, which increased import demand and tariff revenue. When supported by fiscal reforms the tight monetary policy now became more successful. The overall rate of inflation declined from 28.8 per cent in 1994 to only about 1.6 per cent in 1995. This dramatic fall was also due to a stabilised exchange rate and increased supply of maize, following a bumper harvest in 1994.

Figure 11.1 illustrates the extraordinary events that took place between 1992 and 1993. There was a sharp movement towards the north-east. As a reflection of monetary contraction and currency depreciation the *L'L'*-curve moved upwards to *LL*, while the effects of fiscal laxity moved the *NN*-curve to the right (*N'N'*). There were also the added effects of terms-of-trade improvements. It is worth noting that despite the tight monetary policy, fiscal adjustment was not achieved during 1994. The improvement in the external balance was mainly due to a reduction of private expenditures. It was not until 1995 that the fiscal situation was brought under control, with movement to the left of the *N'N'*-curve (to *N"N"*). The movement to the south-west in Figure 11.1 is also partly a reflection of the temporary increase in import tariffs, part of the fiscal adjustment. Although the government was quite successful in stabilising the economy, the choice of higher protection as a means of raising revenue had a negative impact on export performance. The incentives achieved during 1993–94 were thus eroded. The developments during 1993–95 had a major impact at the sectoral level and particularly within the manufacturing industry. In the following section we attempt to supply some survey evidence of the impact of liberalisation and structural adjustment on the manufacturing sector in Kenya.

11.3 Impact of adjustment: survey evidence

The survey covered food, metal, textiles and wood sub-sectors. A sample of some 224 firms was selected and surveyed once every year from 1993 to 1995 using a detailed questionnaire. The section shows the weighted survey results from sampled firms in relation to their perception and experiences of the adjustment policies pursued. The respondents were asked to show whether some business factors had improved, worsened or not changed at all in the 1994–95 period of the survey. These business factors included export opportunities, access to foreign exchange, cost of foreign exchange, competition from imports or from local firms, product demand, utility prices, and infrastructure provision. They were then asked to rank for each business factor the severity of the problem, using 1 for not being a problem and 5 for being a severe problem. The results are reported for the 1994 and 1995 surveys and (when available) for the 1993 survey.

11.3.1 Export opportunities

The results indicate that export opportunities for all the sampled firms had improved with liberalisation of the economy over this period. On the whole, very few firms indicated that export opportunities were a problem in 1994 (see Table 11.2). In 1994, around 90 per cent of the firms reported that they had no problem with export opportunities. However, during 1995 (see Table 11.3) there were an increasing number of firms reporting some problems. The textile sector was particularly concerned, and here about 37 per cent of the companies were complaining. The major problem was that restrictive quotas were imposed on Kenya's textile exports to the USA.

11.3.2 Access to foreign exchange

Foreign exchange availability had been a problem in Kenya for some time before the liberalisation of foreign exchange transactions. This was because

Table 11.2 Export opportunities, 1994 (%)

	Food	*Wood*	*Textiles*	*Metal*	*All*	*No. of firms*
No change	62.6	84.3	78.2	84.5	79.1	75
Improved	36.2	15.6	21.5	14.6	20.3	19
Worsened	1.2	0.1	0.2	0.8	0.6	1
Severity						
Not a problem	93.6	99.3	79.8	87.2	89.5	82
Slight problem	2.9	0.2	19.3	1.1	5.8	6
Moderate	2.0	0.3	0.5	11.3	4.2	4
Large	0.5	0.1	0.0	0.3	0.2	1
Severe	1.0	0.1	0.5	0.1	0.3	1

Source: RPED Survey, 1993–95.

Table 11.3 Export opportunities, 1995 (%)

	Food	*Wood*	*Textiles*	*Metal*	*All*	*No. of firms*
No change	98.4	62.3	95.2	92.5	87.6	94
Improved	1.5	37.5	4.1	5.0	11.5	12
Worsened	0.1	0.2	0.7	2.6	0.9	1
Severity						
Not a problem	85.4	74.6	62.8	85.7	79.0	89
Slight problem	0.3	12.4	0.7	1.3	3.4	4
Moderate	0.1	0.4	0.0	0.9	0.4	1
Large	13.9	12.3	0.7	2.6	8.2	10
Severe	0.3	0.2	35.8	9.5	9.0	11

Source: RPED Survey, 1993–95.

the authorities used trade policies to protect reserves and at the same time generate fiscal revenue. One strong indication from the survey results is that access to foreign exchange had improved and that majority of the firms did not find it a problem any more. The proportion of the firms' responses show that about 88 per cent indicated that foreign exchange availability was no longer a problem (See Table 11.4). Individual sub-sectors recorded higher proportions than the average. For example, the metal sub-sector in the sample showed that 97 per cent of the firms had no problem with foreign exchange availability.

In the following years of the survey, and as the firms learnt to operate in a liberalised economy, access to foreign exchange became less of a problem to their operations (see Tables 11.5 and 11.6). These results tend to confirm and support the basic proposition that the liberalisation process and removal of controls would solve the problem of constrained supply of foreign exchange.

11.3.3 Cost of foreign exchange

In addition to their perception in relation to access to foreign exchange the firms were also asked to indicate their experiences with the cost of foreign

Table 11.4 Access to foreign exchange, 1993 (%)

Severity	*Food*	*Wood*	*Textiles*	*Metal*	*All*	*No. of firms*
Not a problem	78.5	87.9	83.2	97.3	87.9	194
Slight problem	0.3	0.0	0.0	0.1	0.1	1
Moderate	13.0	5.8	15.4	0.4	8.0	18
Large	0.5	0.0	0.0	0.2	0.1	1
Severe	7.7	6.2	1.5	2.0	3.9	4

Source: RPED Survey, 1993–95.

Table 11.5 Access to foreign exchange, 1994 (%)

	Food	*Wood*	*Textiles*	*Metal*	*All*	*No. of firms*
No change	41.2	69.7	95.8	82.7	75.2	69
Improved	58.0	30.3	4.2	16.9	24.6	23
Worsened	0.8	0.0	0.0	0.4	0.2	1
Severity						
Not a problem	100.0	99.7	81.1	98.9	94.5	84
Slight problem	0.0	0.2	18.7	1.1	5.3	5
Moderate	0.0	0.0	0.2	0.0	0.1	1
Large	0.0	0.1	0.0	0.0	0.1	1
Severe	0.0	0.0	0.0	0.0	0.0	0

Source: RPED Survey, 1993–95.

exchange. As discussed in the second part of the chapter, the survey period 1993–95 was the period when the exchange rate was highly unstable. In 1993 there were speculative attacks and this led to a depreciation of the currency. However, this trend was reversed towards the end of 1994 and early 1995 with a persistent appreciation. Part of this appreciation emanated from an accumulation of foreign exchange due to short-term capital inflows.

On average, in 1994, 15 per cent of the respondents felt that the cost of foreign exchange had become cheaper, 76 per cent indicated that there was no change while 9 per cent felt that the situation had worsened: that is, cost of foreign exchange had escalated. Moreover, in both 1994 and 1995, the responses from the firms indicate that the cost of foreign exchange was no longer a problem (Tables 11.7 and 11.8). Thus in 1994 and 1995, the firms had learnt to live in a liberalised environment and the cost of foreign exchange was no longer a problem. But in 1995, there were some indications that cost of foreign exchange was a moderate problem for about 14 per cent

Table 11.6 Access to foreign exchange, 1995 (%)

	Food	*Wood*	*Textiles*	*Metal*	*All*	*No. of firms*
No change	64.5	73.5	74.4	61.5	68.0	69
Improved	19.7	26.5	25.2	38.5	26.8	27
Worsened	15.8	0.0	0.4	0.0	5.1	5
Severity						
Not a problem	99.9	87.8	81.7	88.4	91.1	98
Slight problem	0.1	0.1	0.4	0.0	0.1	1
Moderate	0.0	0.0	17.9	11.6	5.9	7
Large	0.0	12.1	0.0	0.0	2.9	4
Severe	0.0	0.0	0.0	0.0	0.0	0

Source: RPED Survey, 1993–95.

Table 11.7 Cost of foreign exchange, 1994 (%)

	Food	*Wood*	*Textiles*	*Metal*	*All*	*No. of firms*
No change	44.1	69.7	95.6	82.7	75.7	69
Improved	32.5	15.6	3.0	15.3	15.3	14
Worsened	23.4	14.7	1.4	2.0	9.1	8
Severity						
Not a problem	70.3	99.1	98.4	96.2	93.7	83
Slight problem	4.9	0.5	0.0	0.9	1.1	1
Moderate	0.0	0.1	1.2	2.0	1.0	1
Large	24.8	0.1	0.2	0.5	4.0	4
Severe	0.0	0.2	0.2	0.5	0.2	1

Source: RPED Survey, 1993–95.

Table 11.8 Cost of foreign exchange, 1995 (%)

	Food	*Wood*	*Textiles*	*Metal*	*All*	*No. of firms*
No change	76.6	62.1	94.2	73.9	75.8	71
Improved	19.8	0.6	2.2	19.0	10.6	10
Worsened	3.6	37.4	3.6	7.1	13.6	13
Severity						
Not a problem	97.5	74.5	76.7	89.6	86.4	93
Slight problem	0.8	0.5	2.6	3.2	1.6	2
Moderate	0.9	12.3	19.0	3.0	7.5	8
Large	0.6	12.6	0.3	2.6	3.9	4
Severe	0.3	0.0	1.4	1.6	0.7	1

Source: RPED Survey, 1993–95.

Table 11.9 Competition from imports, 1994 (%)

	Food	*Wood*	*Textiles*	*Metal*	*All*	*No. of firms*
No change	72.7	87.1	64.9	96.2	79.5	92
Improved	19.4	0.0	0.3	1.3	3.1	4
Worsened	7.9	12.9	34.9	2.5	17.4	20
Severity						
Not a problem	86.3	99.4	65.1	96.7	84.5	95
Slight problem	6.6	0.3	0.3	0.8	1.1	1
Moderate	5.6	0.2	1.1	0.8	1.3	1
Large	1.5	0.1	22.0	0.9	8.5	9
Severe	0.0	0.1	11.6	0.8	4.5	5

Source: RPED Survey, 1993–95.

of the firms and 13.6 per cent of the firms indicated an increase in the cost of foreign exchange. These responses are consistent with the macroeconomic conditions at the time. The exchange rate had depreciated drastically with the expectations of an aid freeze in the consultative Group Meeting of Donors in Paris in 1995. The instability of the exchange rate in this period thus created uncertainties regarding the future returns of exports and the valuation of stocks in dollar terms.

11.3.4 Competition from imports

With the continuation of liberalisation process, especially liberalisation of trade and foreign exchange transactions, Kenyan firms were continuously being exposed to competition from imports. From the responses in 1994 and 1995 (See Tables 11.9 and 11.10), a large number of firms had adjusted to competition and responded that competition from imported goods was not a problem. This may be viewed as supporting evidence of the adaptability

Table 11.10 Competition from imports, 1995 (%)

	Food	*Wood*	*Textiles*	*Metal*	*All*	*No. of firms*
No change	97.5	81.5	63.9	75.6	78.7	101
Improved	0.2	9.0	0.7	9.6	5.1	7
Worsened	2.3	9.5	35.4	14.8	16.3	21
Severity						
Not a problem	98.2	99.3	42.4	83.3	80.6	110
Slight problem	0.3	0.0	11.2	1.2	3.3	4
Moderate	0.8	0.4	0.7	10.7	3.0	4
Large	0.4	0.2	11.8	2.7	3.8	5
Severe	0.4	0.2	34.0	2.1	9.3	13

Source: RPED Survey, 1993–95.

of the manufacturing sector to liberalisation in Kenya, but this does not imply that all firms were faring well. In 1995, 9.3 per cent of the firms indicated that competition was a severe problem.

Most of these firms are to be found in the textile sub-sector (34 per cent) and, also in this sub-sector, over half of the firms interviewed indicated that competition was a problem to their operations. This supports the fact that the textile industry was undergoing a difficult period trying to compete with second-hand clothing. Many of the medium and small-scale firms in this sub-sector closed down productive lines to establish a commercial line with imported fabrics in line with domestic demand.

11.3.5 Competition from local firms

It was anticipated that with the liberalisation of the economy, competition would become more intense as a more level playing field was created. For majority of the firms, there was no noticeable change, but for specific sub-sectors, competition had worsened. For the food sub-sector in 1994 (See Table 11.11), about 66 per cent indicated no change in competition from local firms, while about 32 per cent indicated that competition had worsened, and was actually a problem to the firms. Only 35 per cent of the firms in the food sub-sector indicated that competition was not a problem. In the wood sub-sector, 47 per cent claimed that competition from local firms had not changed while 53 per cent indicated that it had worsened and approximately 60 per cent saw it as a problem. In the textile and metal sub-sectors, 47 per cent and 56 per cent of the firms indicated no change in competition, while 44 per cent and 31 per cent felt that competition from local firms had worsened. However, 27 per cent and 24 per cent (respectively) of the firms in these two sub-sectors considered competition from local firms to be a large problem. As the survey continued into 1995 (see Table 11.12) problems with domestic competition escalated. On the whole,

Table 11.11 Competition from domestic industries, 1994 (%)

	Food	*Wood*	*Textiles*	*Metal*	*All*	*No. of firms*
No change	66.0	46.8	46.7	56.2	52.3	95
Improved	2.0	0.0	9.0	12.7	6.5	12
Worsened	32.1	53.2	44.3	31.1	41.2	75
Severity						
Not a problem	35.3	40.2	43.1	43.8	40.7	71
Slight problem	6.3	6.7	9.7	6.4	7.2	13
Moderate	29.9	39.5	9.0	24.6	25.2	45
Large	28.5	13.5	26.6	24.4	22.4	40
Severe	0.0	0.1	11.6	0.8	4.5	5

Source: RPED Survey, 1993–95.

Table 11.12 Competition from domestic industries, 1995 (%)

	Food	*Wood*	*Textiles*	*Metal*	*All*	*No. of firms*
No change	55.2	50.0	78.6	62.7	59.9	116
Improved	0.4	10.0	0.6	13.9	6.6	13
Worsened	44.4	40.1	20.8	23.4	33.5	65
Severity						
Not a problem	44.0	25.2	39.0	35.8	35.1	68
Slight problem	32.8	24.8	20.3	2.2	20.8	40
Moderate	0.9	15.1	20.1	14.6	12.5	24
Large	0.4	15.0	10.4	27.6	13.3	26
Severe	21.9	19.8	10.1	19.7	18.4	36

Source: RPED Survey, 1993–95.

18 per cent in 1995 indicated that domestic competition had become severe while only 35 per cent of the firms considered that domestic competition was not a problem to their operations. Interestingly the severity of the problem dropped in the textile sector, while it increased in the other sectors.

11.3.6 Access to imported raw materials

The results support the proposition that liberalisation released the market and economic activities from frustrated demand under the control regime; firms had acquired a tendency to stockpile raw material and inventories to tide them over periods of supply constraints. This would be taken as a positive pointer to the success of liberalisation. When asked whether firms considered access to imported raw materials a problem, majority of the firms indicated that it was not a problem (79 per cent: Table 11.13). These results improved in the 1995 surveys (Table 11.14) where 89 per cent of the firms indicated no problem.

Table 11.13 Access to imported raw materials, 1994 (%)

	Food	*Wood*	*Textiles*	*Metal*	*All*	*No. of firms*
No change	70.3	74.7	70.7	77.2	73.7	81
Improved	28.0	25.3	1.9	3.4	12.0	13
Worsened	1.7	0.0	27.4	19.3	14.3	16
Severity						
Not a problem	97.5	87.8	58.6	82.0	78.8	88
Slight problem	0.0	0.1	0.4	8.9	3.0	3
Moderate	1.3	12.2	14.1	0.3	7.4	8
Large	0.5	0.0	27.0	8.9	10.7	12
Severe	0.8	0.0	0.0	0.0	0.1	1

Source: RPED Survey, 1993–95.

Table 11.14 Access to imported raw materials, 1995 (%)

	Food	*Wood*	*Textiles*	*Metal*	*All*	*No. of firms*
No change	84.0	74.3	74.7	70.4	76.6	92
Improved	15.9	1.2	25.1	7.7	13.3	16
Worsened	0.1	24.5	0.2	21.8	10.1	12
Severity						
Not a problem	99.5	56.5	98.2	96.5	89.0	115
Slight problem	0.4	10.8	0.7	1.8	3.1	4
Moderate	0.0	21.6	0.7	1.7	5.3	7
Large	0.1	10.9	0.4	0.0	2.6	3
Severe	0.0	0.1	0.0	0.0	0.0	1

Source: RPED Survey, 1993–95.

11.3.7 Cost of imported raw materials

With the sudden devaluation of the Kenyan shilling in 1993, it was felt that the cost of imported raw materials had steadily risen and could have been a hindrance to enterprise development or operations. In addition, instability of the exchange rate in 1993–95 could have generated excessive costs due to the pass-through effects from the exchange rate via import prices to domestic production activities. The responses for 1994 (Table 11.15) showed that there was no change in the cost of imported raw materials for about 64 per cent of the firms and 80 per cent for the sampled firms in 1995 (Table 11.16). In addition, 68 per cent of the firms did not consider this as a problem in 1994 and this figure was 75 per cent in 1995. The proportion of the firms that considered the cost of raw materials as a severe problem declined from 6 per cent in 1994 to 3 per cent in 1995.

Table 11.15 Cost of imported raw materials, 1994 (%)

	Food	*Wood*	*Textiles*	*Metal*	*All*	*No. of firms*
No change	69.4	55.5	65.2	67.1	63.6	69
Improved	5.5	0.2	1.0	2.3	1.8	2
Worsened	25.1	44.3	33.8	30.6	34.6	37
Severity						
Not a problem	73.8	77.7	56.7	68.0	68.2	77
Slight problem	2.1	0.1	13.8	10.5	7.4	8
Moderate	20.6	0.0	15.6	1.4	7.7	9
Large	3.4	11.1	13.8	9.9	10.5	12
Severe	0.0	11.1	0.2	10.2	6.2	7

Source: RPED Survey, 1993–95.

Table 11.16 Cost of imported raw materials, 1995 (%)

	Food	*Wood*	*Textiles*	*Metal*	*All*	*No. of firms*
No change	98.1	86.2	57.7	69.7	79.6	93
Improved	0.5	1.0	0.3	4.0	1.3	2
Worsened	1.5	12.8	41.9	26.4	19.1	22
Severity						
Not a problem	99.0	62.4	43.6	81.2	75.0	92
Slight problem	0.3	0.2	13.9	2.0	3.7	4
Moderate	0.3	12.6	14.8	2.6	6.7	7
Large	0.2	12.4	27.7	13.2	11.8	12
Severe	0.1	12.4	0.0	1.0	2.8	3

Source: RPED Survey, 1993–95.

11.3.8 Opportunities to buy foreign machinery and equipment

The positive outcomes of the reform process have been the removal of the foreign exchange constraints as well as the simplification of import procedure. This made it easier to buy foreign machinery and equipment, as was realised during 1994 and 1995. It was not perceived as a problem by 88 per cent of the firms interviewed in 1994, and 95 per cent of those firms interviewed in 1995 (See Tables 11.17 and 11.18). Few firms reported that it was a severe problem and even fewer firms showed that the problem had worsened over time.

This points to the fact that the liberalisation and adjustment process had allowed the firms to base their investment decisions, such as acquiring new capital equipment, on the determining fundamentals rather than be controlled via import licensing or availability and allocation of foreign exchange. This is of major importance, as inflows of capital goods produced

Table 11.17 Opportunities to buy foreign machinery and equipment, 1994 (%)

	Food	*Wood*	*Textiles*	*Metal*	*All*	*No. of firms*
No Change	80.2	59.8	96.1	85.5	78.6	80.0
Improved	18.9	30.4	3.2	14.2	17.7	18
Worsened	1.0	9.8	0.7	0.3	3.7	4
Severity						
Not a problem	95.5	80.1	80.8	98.7	87.6	86
Slight problem	1.3	0.1	0.0	1.2	0.5	1
Moderate	1.9	0.1	0.2	0.1	0.3	1
Large	1.3	19.6	19.0	0.0	11.6	11
Severe	0.0	0.1	0.0	0.0	0.0	1

Source: RPED Survey, 1993–95.

Table 11.18 Opportunities to buy foreign machinery and equipment, 1995 (%)

	Food	*Wood*	*Textiles*	*Metal*	*All*	*No. of firms*
No change	80.0	60.7	77.8	47.5	64.4	85
Improved	19.8	39.3	21.9	52.5	35.5	47
Worsened	0.1	0.0	0.3	0.0	0.1	1
Severity						
Not a problem	99.8	92.2	98.6	92.5	95.4	138
Slight problem	0.2	7.8	0.7	7.5	4.4	6
Moderate	0.0	0.1	0.0	0.0	0.1	1
Large	0.0	0.0	0.4	0.0	0.1	1
Severe	0.0	0.0	0.3	0.0	0.1	1

Source: RPED Survey, 1993–95.

in the industrialised world embody ideas which can have important intangible effects on economic growth (De Long and Summers, 1993).

11.3.9 Business support services

In 1993 (Table 11.19), about 25 per cent of the firms indicated that lack of business support services was a severe problem to the operations. This proportion declined to 5 per cent in 1994 (Table 11.20) and further to 4 per cent in 1995 (Table 11.21). But on average, slightly below half of the firms in the sample in 1994 and 1995 indicated that lack of business support services was a problem, ranging from slight to severe.

This evidence corroborates the impression of declining business support infrastructures in Kenya over this period. Most of the firms had to provide their own support services and some received, support from their trade associations, while public sector provision of business support services was in a poor state or totally absent. This is an area of contradiction where

Table 11.19 Business support services, 1993 (%)

Severity	*Food*	*Wood*	*Textiles*	*Metal*	*All*	*No. of firms*
Not a problem	67.9	47.3	79.4	75.2	67.4	151
Slight problem	1.8	11.6	0.0	0.4	3.7	4
Moderate	3.2	0.3	0.5	6.7	2.5	3
Large	1.2	0.1	5.3	0.2	1.8	2
Severe	25.9	40.8	14.8	17.5	24.6	25

Source: RPED Survey, 1993–95.

Table 11.20 Business support services, 1994 (%)

	Food	*Wood*	*Textiles*	*Metal*	*All*	*No. of firms*
No change	72.2	87.3	91.0	68.9	81.5	142
Improved	14.4	6.4	8.9	9.2	9.0	16
Worsened	13.4	6.3	0.1	21.9	9.5	16.6
Severity						
Not a problem	54.9	47.3	80.2	54.2	59.4	97
Slight problem	17.2	13.1	0.1	15.0	10.6	17
Moderate	1.4	19.8	10.0	15.6	13.4	22
Large	26.5	13.1	0.1	15.2	11.9	20
Severe	0.0	6.7	9.6	0.0	4.8	5

Source: RPED Survey, 1993–95.

Table 11.21 Business support services, 1995 (%)

	Food	*Wood*	*Textiles*	*Metal*	*All*	*No. of firms*
No change	85.4	80.2	84.1	83.4	83.0	125
Improved	14.3	19.8	0.3	8.6	12.3	18
Worsened	0.3	0.1	15.6	8.0	4.8	7
Severity						
Not a problem	70.4	27.4	35.1	61.4	48.6	73
Slight problem	15.0	13.3	2.5	7.5	10.4	16
Moderate	14.3	39.4	46.6	23.1	29.8	45
Large	0.3	6.8	15.8	7.2	6.7	10
Severe	0.1	13.1	0.0	0.8	4.4	7

Source: RPED Survey, 1993–95.

government institutions have failed to provide the required public services to support the manufacturing sector in Kenya. Perhaps the outcome partly reflects the fact that majority of the firms provided their own business sup port services when those provided by the public sector or other agent fell short of their requirements. Viewed this way, it partly answers the apparent contradiction in these results that the majority of the firms indicated that business support services had not changed in the adjustment period, yet the majority of the firms indicated that it was not a problem to them.

11.3.10 Utility prices

One area which was covered in the survey was utility prices. In 1993, 70 per cent of the firms indicated that utility prices were not a problem (Table 11.22). This outcome was reverses in the following years when 72 per cent and 69 per cent of the firms indicated that utility prices had worsened (Tables 11.23 and 11.24). This is further corroborated by the fact that only a small proportion of firms (28 per cent and 31 per cent respectively for 1994 and 1995) indicated that utility prices were not a problem to their operations. A majority of the firms seem to have suffered from escalating utility prices which affected the competitiveness of their products.

The conclusion that can be drawn regarding utility prices in the adjustment period is that it did not support enterprise growth, but rather hindered it. One of the problems here is that there is a combination of high utility prices and poor services. This provides yet another contradiction as

Table 11.22 Utility prices, 1993 (%)

Severity	*Food*	*Wood*	*Textiles*	*Metal*	*All*	*No. of firms*
Not a problem	53.1	58.6	88.1	69.3	69.5	156
Slight problem	14.1	6.0	5.2	0.3	5.3	12
Moderate	3.1	12.0	0.9	17.8	9.1	21
Large	2.8	0.1	5.5	0.8	2.3	5
Severe	26.8	23.3	0.4	11.8	13.8	31

Source: RPED Survey, 1993–95.

Table 11.23 Utility prices, 1994 (%)

	Food	*Wood*	*Textiles*	*Metal*	*All*	*No. of firms*
No change	1.3	24.6	53.1	24.2	28.0	46
Improved	98.7	75.4	46.9	75.8	72.0	117
Worsened	0.0	0.0	0.0	0.0	0.0	0.0
Severity						
Not a problem	1.3	32.7	32.3	34.3	28.2	47
Slight problem	1.5	8.3	42.2	22.6	20.6	34
Moderate	17.9	8.5	22.4	18.4	16.8	28
Large	40.2	33.4	2.4	18.3	21.5	36
Severe	39.2	17.1	0.8	6.4	12.9	21

Source: RPED Survey, 1993–95.

Table 11.24 Utility prices, 1995 (%)

	Food	*Wood*	*Textiles*	*Metal*	*All*	*No. of firms*
No change	11.2	35.2	47.7	26.8	28.4	49
Improved	10.7	0.0	0.0	0.0	3.0	5
Worsened	78.1	64.8	52.3	73.2	68.5	117
Severity						
Not a problem	32.8	14.1	58.8	26.4	31.0	53
Slight problem	21.6	21.2	1.4	19.0	17.0	29
Moderate	22.6	28.8	1.5	20.2	19.7	34
Large	11.6	28.5	37.4	20.6	23.2	40
Severe	11.4	7.4	0.9	13.7	9.0	15

Source: RPED Survey, 1993–95.

Table 11.25 Lack of demand, 1993 (%)

	Food	*Wood*	*Textiles*	*Metal*	*All*	*No. of firms*
Severity						
Not a problem	34.8	41.2	49.3	41.7	42.8	96
Slight problem	0.8	6.0	5.2	6.3	5.1	12
Moderate	26.0	17.6	25.0	23.0	22.4	50
Large	12.6	6.0	5.4	6.2	6.8	15
Severe	25.8	29.2	15.1	22.8	22.8	51

Source: RPED Survey, 1993–95.

liberalisation is supposed to encourage competitive production, but utility prices, mostly from parastatals owned and managed by the government, halt the liberalisation process and competitive production.

11.3.11 Product demand

Following crisis management in the economy during 1993–95, lack of demand became a serious problem for the private sector. Already during 1993 more than half of the firms had a problem (Table 11.25) and it continued during 1994 and 1995 (Tables 11.26 and 11.27). A striking result is that a majority of the firms, 51 per cent, indicated that product demand had fallen in 1994, and 50 per cent saw this as a large or severe problem during 1994. The textile and metal sub-sectors were particular affected during 1994, while the wood sector became negatively affected during 1995.

As discussed in section 11.2, by March of 1993 speculation in foreign exchange was rife. Kenya was in serious danger of a flight from the shilling and increasingly prices were being quoted in dollars. There was also increased domestic currency circulating in the economy, partly due to the

Table 11.26 Lack of demand, 1994 (%)

	Food	*Wood*	*Textiles*	*Metal*	*All*	*No. of firms*
No change	28.9	23.9	17.0	33.1	25.4	50
Improved	51.8	17.7	16.5	22.6	23.3	46
Worsened	19.3	58.5	66.5	44.3	51.3	102
Severity						
Not a problem	67.0	41.1	25.2	29.9	36.9	72
Slight problem	3.6	0.2	1.0	6.0	2.5	5
Moderate	26.7	6.1	0.6	17.6	10.6	21
Large	0.8	34.9	24.8	11.9	20.9	41
Severe	2.0	17.7	48.4	34.6	29.1	57

Source: RPED Survey, 1993–95.

Table 11.27 Lack of demand, 1995 (%)

	Food	*Wood*	*Textiles*	*Metal*	*All*	*No. of firms*
No change	44.2	35.0	48.3	35.9	40.2	78
Improved	44.1	20.3	10.3	32.8	27.0	52
Worsened	11.7	44.8	41.3	31.3	32.8	63
Severity						
Not a problem	66.0	25.2	30.0	38.3	39.3	77
Slight problem	11.1	10.2	9.7	13.2	11.0	22
Moderate	11.6	10.1	28.9	14.3	15.2	30
Large	10.8	19.9	20.4	20.5	17.9	35
Severe	0.5	34.6	11.0	13.7	16.6	33

Source: RPED Survey, 1993–95.

Treasury financing its deficit domestically, while seasonal increases in cash (which coincided with the 1992 elections) put yet more pressure on the domestic prices. In addition, the drought of the 1990/91 continued into 1993, which meant that food prices continued to rise while large amounts of foreign exchange were required for food imports. Furthermore, there were financial scams (involving a few banks) which contributed to increased money supply in the economy and which led to the collapse of these banks. The main preoccupation of the monetary authorities during 1993/94 was therefore to mop up excess liquidity in the economy through aggressive sales of Treasury Bills. Moreover, as foreign reserves were almost depleted in early 1993, a lot of effort was targeted at increasing reserves as well. Even when inflation was tamed in 1994 and 1995, prices did not come down, so that real income declined by over 50 per cent for the public service workers and slightly below 50 per cent for the private sector workers. This had the effect of eroding purchasing power and pushing the recession further, and thus product demand declined drastically for most firms, especially in the textile, metal and wood sub-sectors.

11.3.12 Summary of the adjustment effects

Perhaps not surprisingly, access to credit was the severest problem during all years of survey.[3] This was mainly a result of economic policies implemented to stabilise the economy following severe shocks. Although this had a negative impact on the operations of the private sector, it shows that the effects of the policies trickled down quite fast to the firm level in the Kenyan economy. Lack of demand is another issue related to the tight policy pursued at the time when the survey was conducted, which also had a severe impact on firms' operations. Most of the issues outside firms' control (e.g., utility prices and infrastructure), or else internal re-organisation of the firms ended up being large and severe problems to the firms. Many problems outlined, which were either internal or to which the firm could seek solutions, did not emerge as major problems (for example, cost of raw materials, cost of foreign exchange or business support services, to mention but a few). Institutions define the incentive structure and are also the conduit between the manufacturing sector and government policy thrust and formulation. The identified problems are at the heart of those fundamentals that define and enhance the profitability of domestic production. These factors thus hinder the appropriate supply response of domestic industries and would-be investors. This may perhaps provide some support to the out-turn of the survey results in Kenya and explain why there has been an investment pause in the manufacturing sector and a contraction of the sector in the liberalisation period, thus hindering the adjustment process and supply response of the liberalisation environment.

11.4 Conclusions

Not unexpectedly, the Kenyan adjustment experience has not been smooth. Policy reversals, terms-of-trade shocks and unfavourable movements in aid flows contributed to an uneven performance. In response to the crisis year of 1993, Kenyan authorities gave recovery the highest priority. There has been progress in some areas, and it looks like a fragile recovery has been accomplished. The fragility stems mainly from uncertainty about the political process, fear of policy reversals, and corruption in the civil service, all of which militate against both foreign investment and domestic investment. The tight monetary policy regime was quite successful in stabilising the economy; money supply growth was on target; and inflation came down drastically during 1995. However, the austerity measures pushed the economy into a recession during the survey period. Lack of demand became a major problem to firms' operations. Moreover, the high interest rates prevailing slowed down investment activities and also became a severe restraint on firms' operations. Thus, a stable macroeconomic environment is a prerequisite for private-sector-led growth.

There are indications that the liberalised environment has had negative effects in some sectors. Although the reform process will continue to hurt some sectors, it is not desirable to move back to a controlled regime. Most actors in the private sector, even within the textile sector, support the liberalised regime, but they argue that strong action has to be taken against corruption, and against lack of transparency in customs procedures.

In the adjustment period, poor infrastructure and escalating utility prices did not support enterprise growth, but rather hindered it. A major problem is that there is a combination of high utility prices and poor services. This provides a severe bottleneck for private sector development. Liberalisation is supposed to encourage competitive production, but utility prices and poor infrastructure halts the liberalisation process and competitive production.

Appendix

Table A11.1 Price indices for Kenya, 1985–95 (1986 = 100)

	1986	*1987*	*1988*	*1989*	*1990*	*1991*	*1992*	*1993*	*1994*	*1995*
P_x	100.0	100.6	109.8	118.9	120.1	130.7	154.7	214.5	252.9	301.4
P_m	100.0	101.5	110.5	118.5	130.0	148.0	164.1	174.8	204.9	251.1
P_n	100.0	108.8	120.3	137.7	166.0	177.4	223.4	277.3	329.5	373.9
P_x/P_m	100.0	99.1	99.4	100.4	92.4	88.3	94.2	122.7	123.4	120.0
P_n/P_m	100.0	107.3	108.9	116.2	127.7	119.9	136.1	158.7	160.8	148.9
P_m/P_n	100.0	93.2	91.9	86.1	78.3	83.4	73.5	63.0	62.2	67.2
P_x/P_n	100.0	92.4	91.3	86.4	72.4	73.7	69.2	77.3	76.7	80.6
TOT	100.0	81.8	85.5	76.2	68.7	78.8	76.9	87.2	97.8	93.1

Notes:
P_x: The GDP deflator for agriculture used as a proxy for prices of exportables.
P_m: The GDP deflator for manufacturing used as a proxy for prices of import substitutes.
P_n: The GDP deflator for building and construction, wholesale and retail trade, transport, storage and communication, hotels and restaurants is used as a proxy for prices of non-tradables prices.
TOT: International terms of trade
Source: Republic of Kenya, *Economic Survey*, various issues.

Table A11.2 The biggest problem, 1993 (%)

	Food	*Wood*	*Textiles*	*Metal*	*All*	*No. of firms*
Ownership regulations	1.0	0.0	0.0	0.0	0.1	1
Taxes	0.4	0.2	5.1	0.0	1.6	4
Government restrictions on activities	12.6	5.8	0.0	5.7	5.0	11

Table A11.2 *continued*

	Food	*Wood*	*Textiles*	*Metal*	*All*	*No. of firms*
Gaining investment benefits	0.0	0.0	0.0	0.1	0.0	1
Labour regulations	0.0	0.1	0.0	5.7	1.6	4
Difficulty in obtaining licences	0.0	0.0	0.1	0.0	0.0	1
Corruption	0.0	0.0	5.1	0.2	1.6	4
Price controls	12.9	0.0	0.0	0.0	1.8	4
Lack of business support services	0.8	0.0	5.1	0.0	1.6	4
Lack of infrastructure	0.4	11.7	4.8	28.4	12.7	29
Utility prices	1.2	5.9	9.7	0.1	4.8	11
Access to credit	13.4	63.9	44.1	23.2	39.8	89
No demand	36.7	6.1	4.8	23.0	14.6	33
Access to foreign exchange	3.3	0.2	0.7	1.3	1.1	2
Competition from imports	12.2	0.0	5.3	0.6	3.4	8
Competition from local firms	–	–	–	–	–	–
Uncertainty about government industry policies	0.3	0.0	0.0	0.0	0.0	0
Lack of skilled labour	0.0	0.1	0.0	0.2	0.1	0
Other	4.4	6.0	10.3	11.6	8.6	9
Security	0.5	0.0	4.9	0.0	1.5	2
Access to imported raw materials	–	–	–	–	–	–
Cost of imported raw materials	–	–	–	–	–	–
Access to domestic raw materials	–	–	–	–	–	–
Cost of domestic raw materials	–	–	–	–	–	–
High interest rates	–	–	–	–	–	–
Inflation	–	–	–	–	–	–
High exchange rate	–	–	–	–	–	–

Table A11.3 The second biggest problem, 1993 (%)

	Food	*Wood*	*Textiles*	*Metal*	*All*	*No. of firms*
Ownership regulations	0.0	0.0	0.0	0.3	0.1	1
Taxes	0.4	0.3	0.0	13.1	3.6	7
Government restrictions on activities	24.9	0.0	0.1	0.0	3.6	7
Gaining investment benefits	0.0	0.0	0.0	0.0	0.0	0
Labour regulations	0.0	0.1	0.2	0.3	0.2	1
Difficulty in obtaining licences	1.8	0.1	16.0	6.6	6.7	14
Corruption	12.3	0.0	5.3	0.1	3.4	7
Price controls	0.5	0.0	0.0	0.1	0.1	1
Lack of business support services	0.8	6.2	10.6	6.8	6.9	14
Lack of infrastructure	1.2	12.5	10.8	19.2	12.2	25
Utility prices	14.0	0.0	0.2	0.3	2.2	4
Access to credit	14.2	31.1	27.3	19.9	24.5	50

Table A11.3 *continued*

	Food	*Wood*	*Textiles*	*Metal*	*All*	*No. of firms*
No demand	0.6	18.6	6.4	19.3	12.6	26
Access to foreign exchange	3.0	0.1	0.6	0.5	0.8	2
Competition from imports	0.0	0.0	5.8	0.4	1.8	4
Competition from local firms	0.6	6.2	0.0	0.0	1.9	4
Uncertainty about government industry policies	0.0	0.1	0.0	0.0	0.0	0
Lack of skilled labour	0.0	0.0	0.1	0.0	0.0	0
Other	25.7	18.5	16.6	6.6	15.8	32
Security	0.0	6.2	0.0	6.5	3.6	7
Access to imported raw materials	–	–	–	–	–	–
Cost of imported raw materials	–	–	–	–	–	–
Access to domestic raw materials	–	–	–	–	–	–
Cost of domestic raw materials	–	–	–	–	–	–
High interest rates	–	–	–	–	–	–
Inflation	–	–	–	–	–	–
High exchange rate	–	–	–	–	–	–

Table A11.4 The third biggest problem, 1993 (%)

	Food	*Wood*	*Textiles*	*Metal*	*All*	*No. of firms*
Ownership regulations	0.0	0.0	0.0	0.0	0.0	0
Taxes	3.8	0.1	0.4	0.9	0.9	1
Government restrictions on activities	0.4	6.7	0.0	0.0	2.2	4
Gaining investment benefits	0.0	0.0	0.2	0.1	0.1	1
Labour regulations	1.2	0.1	0.1	0.3	0.3	1
Difficulty in obtaining licences	1.2	6.6	0.0	6.9	4.4	8
Corruption	0.7	0.0	0.2	0.0	0.1	1
Price controls	0.0	0.0	0.0	0.0	0.0	0
Lack of business support services	0.8	13.2	0.1	14.0	8.6	15
Lack of infrastructure	19.2	26.5	14.9	0.6	15.0	26
Utility prices	1.0	13.2	0.4	7.3	6.7	12
Access to credit	34.4	0.1	0.2	7.0	6.6	11
No demand	0.4	13.4	30.7	20.5	18.1	32
Access to foreign exchange	17.1	6.6	0.0	0.3	4.4	8
Competition from imports	0.7	0.1	29.8	0.4	7.7	13
Competition from local firms	2.1	0.0	0.4	20.5	6.4	11
Uncertainty about government industry policies	0.0	0.0	0.1	0.1	0.1	1
Lack of skilled labour	0.0	0.0	0.2	6.8	2.1	4
Other	0.9	6.7	0.0	13.9	6.4	11

Table A11.4 continued

	Food	*Wood*	*Textiles*	*Metal*	*All*	*No. of firms*
Security	16.4	6.6	22.2	0.3	9.9	17
Access to imported raw materials	–	–	–	–	–	–
Cost of imported raw materials	–	–	–	–	–	–
Access to domestic raw materials	–	–	–	–	–	–
Cost of domestic raw materials	–	–	–	–	–	–
High interest rates	–	–	–	–	–	–
Inflation	–	–	–	–	–	–
High exchange rate	–	–	–	–	–	–

Table A11.5 The biggest problem, 1994 (%)

	Food	*Wood*	*Textiles*	*Metal*	*All*	*No. of firms*
Ownership regulations	0.4	0.0	0.0	0.0	0.1	1
Taxes	0.9	0.1	0.0	0.1	0.2	1
Government restrictions on activities	0.0	0.0	0.1	0.0	0.1	1
Gaining investment benefits	0.0	0.0	0.0	0.0	0.0	0
Labour regulations	0.0	0.0	0.0	0.0	0.0	0
Difficulty in obtaining licences	0.0	0.0	0.0	0.0	0.0	0
Corruption	0.7	0.1	0.1	0.0	0.1	1
Price controls	0.2	0.0	0.0	0.0	0.1	1
Lack of business support services	0.0	5.5	0.0	0.0	1.6	3
Lack of infrastructure	11.3	5.6	0.1	5.9	4.8	10
Utility prices	11.1	0.0	0.0	0.0	1.5	3
Access to credit	23.6	33.1	41.5	32.8	34.2	73
No demand	11.1	27.6	21.6	21.9	22.0	47
Access to foreign exchange	0.0	0.0	0.0	0.1	0.1	1
Competition from imports	0.4	0.0	7.3	0.4	2.3	5
Competition from local firms	1.6	0.0	6.9	0.0	2.3	5
Uncertainty about government industry policies	0.0	0.0	0.0	0.2	0.1	1
Lack of skilled labour	10.4	5.7	0.0	0.3	3.1	7
Other	28.3	22.3	22.4	38.4	27.6	59
Security	–	–	–	–	–	–
Access to imported raw materials	–	–	–	–	–	–
Cost of imported raw materials	–	–	–	–	–	–
Access to domestic raw materials	–	–	–	–	–	–
Cost of domestic raw materials	–	–	–	–	–	–
High interest rates	–	–	–	–	–	–
Inflation	–	–	–	–	–	–
High exchange rate	–	–	–	–	–	–

Table A11.5 The biggest problem, 1994 (%)

	Food	*Wood*	*Textiles*	*Metal*	*All*	*No. of firms*
Ownership regulations	0.4	0.0	0.0	0.0	0.1	1
Taxes	0.9	0.1	0.0	0.1	0.2	1
Government restrictions on activities	0.0	0.0	0.1	0.0	0.1	1
Gaining investment benefits	0.0	0.0	0.0	0.0	0.0	0
Labour regulations	0.0	0.0	0.0	0.0	0.0	0
Difficulty in obtaining licences	0.0	0.0	0.0	0.0	0.0	0
Corruption	0.7	0.1	0.1	0.0	0.1	1
Price controls	0.2	0.0	0.0	0.0	0.1	1
Lack of business support services	0.0	5.5	0.0	0.0	1.6	3
Lack of infrastructure	11.3	5.6	0.1	5.9	4.8	10
Utility prices	11.1	0.0	0.0	0.0	1.5	3
Access to credit	23.6	33.1	41.5	32.8	34.2	73
No demand	11.1	27.6	21.6	21.9	22.0	47
Access to foreign exchange	0.0	0.0	0.0	0.1	0.1	1
Competition from imports	0.4	0.0	7.3	0.4	2.3	5
Competition from local firms	1.6	0.0	6.9	0.0	2.3	5
Uncertainty about government industry policies	0.0	0.0	0.0	0.2	0.1	1
Lack of skilled labour	10.4	5.7	0.0	0.3	3.1	7
Other	28.3	22.3	22.4	38.4	27.6	59
Security	–	–	–	–	–	–
Access to imported raw materials	–	–	–	–	–	–
Cost of imported raw materials	–	–	–	–	–	–
Access to domestic raw materials	–	–	–	–	–	–
Cost of domestic raw materials	–	–	–	–	–	–
High interest rates	–	–	–	–	–	–
Inflation	–	–	–	–	–	–
High exchange rate	–	–	–	–	–	–

Table A11.6 The second biggest problem, 1994 (%)

	Food	*Wood*	*Textiles*	*Metal*	*All*	*No. of firms*
Ownership regulations	0.0	0.0	0.0	0.0	0.0	0.0
Taxes	0.0	0.1	0.1	0.2	0.1	1
Government restrictions on activities	0.4	0.0	0.0	0.1	0.1	1
Gaining investment benefits	0.0	0.0	0.0	0.0	0.0	0
Labour regulations	0.0	0.0	0.2	0.1	0.1	1
Difficulty in obtaining licences	0.0	0.0	0.0	0.0	0.0	0
Corruption	0.7	13.2	0.0	6.2	5.5	10
Price controls	0.0	0.0	0.0	0.0	0.0	0

Table A11.6 *continued*

	Food	*Wood*	*Textiles*	*Metal*	*All*	*No. of firms*
Lack of business support services	10.6	0.1	8.0	0.3	4.0	8
Lack of infrastructure	12.7	13.2	8.3	12.0	11.4	21
Utility prices	2.3	0.2	0.2	0.4	0.6	1
Access to credit	11.3	19.7	16.5	24.1	18.7	35
No demand	0.9	19.8	32.5	6.8	16.9	32
Access to foreign exchange	0.0	0.0	0.0	0.0	0.0	0
Competition from imports	0.0	0.0	8.3	0.2	2.5	5
Competition from local firms	11.1	0.0	8.1	0.0	4.0	8
Uncertainty about government industry policies	0.2	0.0	0.0	0.2	0.1	1
Lack of skilled labour	2.0	0.2	0.4	6.7	2.4	5
Other	47.6	33.5	17.6	42.7	33.6	63
Security	–	–	–	–	–	–
Access to imported raw materials	–	–	–	–	–	–
Cost of imported raw materials	–	–	–	–	–	–
Access to domestic raw materials	–	–	–	–	–	–
Cost of domestic raw materials	–	–	–	–	–	–
High interest rates	–	–	–	–	–	–
Inflation	–	–	–	–	–	–
High exchange rate	–	–	–	–	–	–

Table A11.7 The third biggest problem, 1994 (%)

	Food	*Wood*	*Textiles*	*Metal*	*All*	*No. of firms*
Ownership regulations	0.0	0.0	0.0	0.0	0.0	0
Taxes	1.7	0.1	0.0	0.1	0.3	1
Government restrictions on activities	0.4	0.0	0.0	0.0	0.1	1
Gaining investment benefits	0.0	0.0	0.2	0.0	0.1	1
Labour regulations	0.0	0.1	0.2	0.0	0.1	1
Difficulty in obtaining licences	0.0	7.6	0.2	0.0	2.5	4
Corruption	3.0	0.1	0.2	0.2	0.4	1
Price controls	0.0	0.0	0.0	0.0	0.0	0
Lack of business support services	0.4	15.1	0.0	13.7	9.5	13
Lack of infrastructure	3.0	7.8	0.2	0.3	3.0	4
Utility prices	3.1	15.4	0.9	0.3	5.6	8
Access to credit	21.5	7.7	13.5	14.2	12.7	18
No demand	0.9	7.6	27.5	14.0	13.6	19
Access to foreign exchange	0.0	0.0	0.0	0.0	0.0	0
Competition from imports	0.0	0.1	40.3	0.2	9.5	13
Competition from local firms	21.5	7.7	0.2	0.4	5.0	7

Table A11.7 continued

	Food	*Wood*	*Textiles*	*Metal*	*All*	*No. of firms*
Uncertainty about government industry policies	0.8	0.1	0.2	7.0	2.5	3
Lack of skilled labour	20.2	0.1	0.7	7.9	5.0	7
Other	23.2	30.7	15.9	41.8	30.1	42
Security	–	–	–	–	–	–
Access to imported raw materials	–	–	–	–	–	–
Cost of imported raw materials	–	–	–	–	–	–
Access to domestic raw materials	–	–	–	–	–	–
Cost of domestic raw materials	–	–	–	–	–	–
High interest rates	–	–	–	–	–	–
Inflation	–	–	–	–	–	–
High exchange rate	–	–	–	–	–	–

Table A11.8 The biggest problem, 1995 (%)

	Food	*Wood*	*Textiles*	*Metal*	*All*	*No. of firms*
Ownership regulations	0.0	0.0	0.0	0.0	0.0	0
Taxes	0.0	0.1	0.0	0.2	0.1	1
Government restrictions on activities	0.1	0.0	0.0	0.0	0.1	1
Gaining investment benefits	0.0	0.0	0.0	0.0	0.0	0
Labour regulations	0.0	0.0	0.0	0.0	0.0	0
Difficulty in obtaining licences	0.0	0.0	0.0	0.0	0.0	0
Corruption	0.2	0.1	0.2	0.0	0.1	1
Price controls	0.0	0.0	0.0	0.0	0.0	0
Lack of business support services	0.0	0.0	0.0	0.0	0.0	0
Lack of infrastructure	11.2	10.0	0.0	12.1	8.5	18
Utility prices	0.0	0.0	0.2	0.4	0.1	1
Access to credit	43.2	44.6	55.6	34.4	44.3	91
No demand	11.0	15.1	1.1	12.5	10.4	21
Access to foreign exchange	0.0	0.0	0.0	0.0	0.0	0
Competition from imports	0.1	0.1	1.1	0.8	0.5	1
Competition from local firms	11.0	0.0	0.2	0.4	2.7	6
Uncertainty about government industry policies	0.2	0.0	0.0	0.3	0.1	1
Lack of skilled labour	0.0	0.2	0.0	0.2	0.1	1
Other	12.0	19.9	33.1	12.8	19.3	40
Security	0.1	0.0	0.0	0.5	0.2	1
Access to imported raw materials	0.0	0.0	0.0	0.0	0.0	0
Cost of imported raw materials	0.0	0.0	0.2	0.5	0.2	1

Table A11.8 continued

	Food	*Wood*	*Textiles*	*Metal*	*All*	*No. of firms*
Access to domestic raw materials	0.3	9.9	8.2	17.5	9.1	19
Cost of domestic raw materials	0.0	0.0	0.2	5.6	1.4	3
High interest rates	0.0	0.0	0.2	1.4	0.4	1
Inflation	10.7	0.0	0.0	0.0	2.5	5
High exchange rate	0.1	0.0	0.0	0.3	0.1	1

Table A11.9 The second biggest problem, 1995 (%)

	Food	*Wood*	*Textiles*	*Metal*	*All*	*No. of firms*
Ownership regulations	0.0	0.0	0.0	0.0	0.0	0
Taxes	0.0	0.0	0.2	0.2	0.1	1
Government restrictions on activities	0.1	5.2	0.0	0.3	1.8	3
Gaining investment benefits	0.0	0.0	0.0	0.0	0.0	0
Labour regulations	0.0	0.0	0.0	0.0	0.0	0
Difficulty in obtaining licences	0.0	0.0	0.0	0.2	0.1	1
Corruption	0.2	5.2	0.6	0.6	2.1	4
Price controls	0.0	0.0	0.0	0.0	0.0	0
Lack of business support services	0.1	0.1	10.4	6.0	3.5	6
Lack of infrastructure	24.6	10.5	0.8	18.4	14.0	26
Utility prices	24.2	0.0	0.0	0.9	5.9	11
Access to credit	12.1	10.4	0.5	12.6	9.4	17
No demand	12.1	10.4	21.2	6.8	11.9	22
Access to foreign exchange	0.0	0.0	0.0	0.0	0.0	0
Competition from imports	0.2	0.1	10.7	0.8	2.3	4
Competition from local firms	0.3	5.2	0.2	0.6	2.0	4
Uncertainty about government industry policies	0.2	0.0	0.0	0.1	0.1	1
Lack of skilled labour	0.0	0.1	0.3	6.6	1.7	3
Other	13.2	26.3	22.5	31.7	23.9	44
Security	0.1	0.1	0.2	0.0	0.1	1
Access to imported raw materials	0.0	0.0	0.2	0.0	0.1	1
Cost of imported raw materials	0.0	0.1	11.0	0.0	2.1	4
Access to domestic raw materials	12.3	10.6	10.4	0.6	8.4	16
Cost of domestic raw materials	0.1	10.4	10.4	6.0	6.9	13
High interest rates	0.1	5.2	0.4	1.6	2.2	4
Inflation	0.0	0.0	0.0	6.0	1.5	3
High exchange rate	0.1	0.0	0.0	0.0	0.1	1

Table A11.10 The third biggest problem, 1995 (%)

	Food	*Wood*	*Textiles*	*Metal*	*All*	*No. of firms*
Ownership regulations	0.0	0.0	0.0	0.0	0.0	0
Taxes	0.2	0.0	0.0	0.3	0.1	1
Government restrictions on activities	0.0	0.2	0.0	0.3	0.1	1
Gaining investment benefits	0.1	0.0	0.0	0.0	0.1	1
Labour regulations	0.1	0.0	0.0	0.0	0.1	1
Difficulty in obtaining licences	0.0	7.0	0.0	0.4	2.4	3
Corruption	0.1	7.3	0.5	16.1	6.6	9
Price controls	0.0	0.0	0.0	0.0	0.0	0
Lack of business support services	0.0	14.0	23.5	0.0	9.8	13
Lack of infrastructure	0.7	7.2	0.9	2.0	3.2	4
Utility prices	0.4	7.0	0.5	0.5	2.6	4
Access to credit	0.4	14.0	11.7	8.9	9.5	13
No demand	19.2	7.3	0.6	0.0	6.3	9
Access to foreign exchange	0.0	7.0	0.0	0.0	2.3	3
Competition from imports	0.0	0.0	12.0	0.0	2.7	4
Competition from local firms	0.3	7.3	12.4	23.9	11.3	15
Uncertainty about government industry policies	0.0	0.1	0.4	0.0	0.1	1
Lack of skilled labour	0.0	0.0	12.2	9.7	5.2	7
Other	58.9	7.2	13.1	27.4	23.8	33
Security	0.1	7.0	12.0	0.9	5.2	7
Access to imported raw materials	0.0	0.0	0.0	0.0	0.0	0
Cost of imported raw materials	0.0	0.0	0.0	0.0	0.0	0
Access to domestic raw materials	19.3	7.1	0.0	8.0	8.1	11
Cost of domestic raw materials	0.1	0.1	0.0	0.0	0.1	1
High interest rates	0.0	0.0	0.2	1.0	0.3	1
Inflation	0.0	0.0	0.0	0.0	0.0	0
High exchange rate	0.1	0.1	0.0	0.7	0.2	1

Notes

1. *NN* is a locus of relative prices, which ensures equilibrium in the non-tradable market. The *NN* locus is drawn on the assumption that real income (y) and real money balances (M/P) are constant. Changes in any of these will shift the curve. The *NN* curve can also shift in response to movements of capital (K_n) into or out of the sector. The *LL* locus denotes values of relative prices which, assuming real income (y), asset demand (A) and money supply expressed in foreign currency (M_s/e) to be constant, lead to equilibrium in the money market. When the price level is too high but below *LL*, the money market is in excess demand, while above it, the market is in excess supply. If money supply were to increase (caused endogenously by a balance of payments surplus or exogenously through discretionary policy), the *LL* curve shifts downwards. See Bevan *et al.* (1990) for further details.
2. The depreciation of the currency in 1993 reflected expansion of money supply and the move towards liberalisation in the goods and foreign exchange markets.

The most spectacular increase in money supply came from what has been referred to as 'election money', used to fund election expenses of the incumbent party.

3. A summary of problem rankings is shown in the Appendix Tables A11.2–11.10.

References

Bevan, D., Collier, P. and Gunning, J.W. with Bigsten, A. and Horsnell, P. (1990), *Controlled Open Economies – A Neoclassical Approach to Structuralism*, Clarendon Press, Oxford.

De Long, J.B. and Summers, L. H. (1993), 'How Strongly do Developing Economies Benefit from Equipment Investment', *Journal of Monetary Economics* 32.

Reinikka, R. (1994), 'How to Identify Trade Liberalization Episodes: An Empirical Study on Kenya', Working Paper 94.10, Institute of Economics and Statistics, University of Oxford.

Republic of Kenya (various issues), *Economic Survey*, Central Bureau of Statistics, Office of the Vice-President and Ministry of Planning and National Development, Nairobi.

Swamy, G. (1994), 'Kenya: Structural Adjustment in the 1980s', World Bank Policy Research Working Paper 1238, Washington, DC.

12
Infrastructure and Enterprise Performance

Steve Kayizzi-Mugerwa and Peter Kimuyu

12.1 Introduction

Kenya had by the 1960s built a fairly modern infrastructure, which helped to consolidate its position as East Africa's commercial centre. The capital city, Nairobi, and other major towns are the hubs of the main regional road, rail and air links. Nairobi is also the hub of telecommunications within the region, while the port of Mombasa is the gateway for many countries in the interior. Since infrastructure and related networks were considered strategic for national development their provision became the responsibility of the public sector. In subsequent decades, however, a combination of population pressure, poor maintenance due to inadequate budgetary outlays, and ineffective macroeconomic policies began to erode the quality of infrastructure provision. From the mid-1970s onwards, the government became increasingly unable to meet the infrastructure needs of the economy. There then followed a sharp regression in infrastructure service delivery and quality during the 1980s and 1990s (Nalo, 1993; Republic of Kenya, 1995). But while the infrastructure's deterioration and its negative impact on the economy are not disputed, it is less clear how firms and sectors have been affected and which factors have determined their response to the deficient infrastructure. Likewise, the effects of poor infrastructure provision on firm investment and factor inputs invite investigation.

Since the early 1980s, Kenya has been implementing structural adjustment policies, a major goal being to increase investment and to encourage the development of the private sector. In a more liberal economic environment, investors, both domestic and foreign, have a much broader choice of where to locate production and other activities. Since a better infrastructure lowers production costs, increases market access and generally raises social welfare, investors prefer, other factors remaining equal, to invest in regions or countries where the supply of infrastructure services is good (Wheeler and Mody, 1992).

In this chapter, we look at infrastructure provision and its effect on manufacturing activities in Kenya. The second section provides an overview of the stock of infrastructure at the national level. This is followed, in the third section, by a review of recent studies on the relationship between infrastructure and enterprise development, including firm responses to deficient provision. Sections 12.4, 12.5 and 12.6 comprise the empirical parts of the chapter. In section 12.4, we discuss the responses of Kenyan firms to inadequate infrastructure provision, based on a number of tabular results. In section 12.5, these issues are explored further with a focus on the amenity effects of infrastructure provision. Section 12.6 looks at the factors that determine the incidence of the self-provision of infrastructure. The chapter is concluded in section 12.7.

12.2 The stock of Kenya's infrastructure

Kenya's road sector is fairly well developed and is the most readily used mode for freight and passenger transport. Although there are attempts to put the management of the entire network under the recently established Kenya Roads Board, a large proportion of the network is, however, 'unclassified' and its maintenance is still the responsibility of poorly funded local authorities or poorly coordinated efforts by line ministries.[1] The 'classified' component of the road network is under the Ministry of Public Works and Housing and covers the main links from Mombasa through Nairobi to the borders with neighbouring countries. The latter are heavily trafficked, with high costs for repair and rehabilitation. Recent years have only seen marginal extensions, while flooding and other weather-related disasters have done considerable damage to roads.

On the whole, paucity of resources has reduced the emphasis on modern infrastructure development, especially in urban areas, to maintenance of the existing network. With donor support, the development of an all-weather road system in the rural areas, in order to incorporate rural dwellers into the modern economy, is being emphasised. This will involve selective upgrading of roads to bitumen standard, and intensive use of labour-based methods for road construction (see World Bank, 1990). In real terms, however, funds for road construction and maintenance declined by 60 per cent between fiscal year 1988/89 and fiscal year 1993/94 (Kenya Association of Manufacturers, 1996). Though much lauded, the Road Maintenance Levy Fund started by the government has yet to generate a reasonable portion of the funds required.

Table 12.1 presents figures on Kenya's 'classified' road network by province. Though comprising a total of over 63 000 km, the condition and standard of the network varied substantially between regions. For the paved parts of the network, close to 9000 km, only 30 per cent were considered to be in good condition, while 20 per cent were considered to be in

Table 12.1 Kenya: road length and surface type for the 'classified' network, 1995

Province	*Total length (km)*	*Bitumen (%)*	*Gravel (%)*	*Earth (%)*
Nairobi	383.7	92	7.97	0.03
Central	7 761.2	25	43.6	31.4
Coast	5 658.9	13	28.8	58.2
Eastern	13 010	8.4	37	54.6
North-eastern	4 854.6	3	13.5	83.5
Western	4 048.7	10	61.3	28.7
Rift Valley	20 203.8	15.6	47.3	37.1
Nyanza	7 225.5	10.3	52	37.7
Total	63 146.4	13.5	41.7	44.8

Source: The Kenya Association of Manufacturers (1996).

critical condition. On the other hand, of the remaining 54 000 km of earth and gravel roads, mainly in the rural areas, only 12 per cent were considered to be in good condition. In some urban areas, including Nairobi, portions of the roads had degenerated considerably, rendering heavy transport close to impossible.

Closely related to the road network is the railway system. The railway played a crucial role in the opening-up of the East African hinterland. However, railways have seen little development in the post-independence years. There have been serious lapses in maintenance, leading to derailments and the collapse of bridges in the different parts of the country. The railway system is at the moment quite inefficient and, though still potentially cheaper than other forms of transport, has been out-competed by the privately owned haulage companies.

Kenya's main port is at Mombasa, while Kisumu is an important port on the shores of Lake Victoria. Both provide services to other countries in the hinterland (Uganda, Rwanda, Burundi, eastern parts of the Democratic Republic of Congo, Southern Sudan, north-eastern Tanzania, Somalia and Ethiopia). The port throughput at Mombasa increased from 6.42 million tonnes in 1990 to about 7.54 tonnes in 1996, but has fluctuated widely in recent years as the volume of relief food through Kenya to neighbouring countries increased (Republic of Kenya, 1994). Traffic is dominated by imports, which account for 70 per cent of the total cargo, with oil accounting for more than half.

However, the quality of services at Mombasa has seen serious deterioration in the past decades. Port users complain about the inadequacy of existing capacity, both in terms of cargo handling and clearance from the port. Under pressure from industry and other users, the government undertook a thorough review of the problems facing the Kenya Ports Authority (KPA) and allied institutions. The review found that both the structure and operation of the KPA were inadequate. But although the government sent out tenders for the management of the container terminal and maintenance of port equipment, even going as far as enlisting foreign expertise, the problems of inefficiency indicated by high pilferage rates and long turnaround rates have persisted.

For many years, Nairobi has been the regional hub for international air traffic. In the past decade, expanding services have contributed to the development of tourism and the export of horticultural products, which are important sources of foreign exchange. The country's sole international airline, Kenya Airways, was privatised, with KLM acquiring the majority stake. Re-capitalised and under new management, Kenya Airways has gone on an aggressive drive to increase market share, opening a number of new routes, notably to West Africa. With the global deregulation of air transport, it is to be surmised that a number of new configurations will emerge in the not-too-distant future, probably involving the now closely cooperating East African countries.

Kenya's posts and telecommunications system was an offshoot of the commonly-owned East African service organisations that preceded independence. The Kenya Posts and Telecommunications Corporation (KPTC) was incorporated in 1977 by an Act of Parliament and has since expanded rapidly. The posts part has, however, performed poorly and has had to be subsidised by the telecommunications. Between 1990 and 1996, postal services grew by some 5 per cent (Republic of Kenya, 1994) which is much lower than the total growth of the population during the period. Deficits in the postal business have meant higher costs for the whole corporation. The KPTC has thus been unable to innovate. Its technology is fairly old, while services for new technologies (such as cellular phones) are expensive. Lack of a performance contract with government has made the KPTC lax in maintenance. As a result, there is a high fault occurrence, coupled with a low fault clearance (Kenya Association of Manufacturers, 1996). The recent move to privatise the telecommunications arm of the KPTC and to introduce other suppliers of telecommunications services promises to introduce more competition and a broader choice for consumers.

With regard to water provision, rapid expansion of the urban population in Kenya, as well as deteriorating access to water in rural areas, especially the arid parts of the country, has put considerable pressure on existing sources. While the national policy objective has been to ensure rapid development in the distribution of water in order to meet the rapidly increasing demand, legislation coordinated towards this goal has been lacking. There is a history of underfunding and previous targets have not been realised. The bulk of the equipment is from the 1960s and there is a serious backlog of repairs and new investment. For example, it was planned that by 1996, 62 per cent of Kenya's population would have access to water through organised water points without indicating how this was to be achieved in practice. It has been estimated that in urban areas only Eldoret has an adequate supply of water (over 125 per cent of demand) while Nairobi meets only 72 per cent of its needs, Mombasa 82, Kisumu 38 and Nakuru 25 per cent (Kenya Association of Manufacturers, 1996).

Kenya has sought to ensure an adequate supply of energy. A number of energy sources are used, although development focuses only on a handful: wood fuel, petroleum and hydroelectric power. Wood provides 70 per cent of the country's total energy requirements. Ease of access has made it the most important energy source, not only in the countryside, but also in the cities. Policy towards the wood sub-sector seeks to ensure adequate wood supply, sustainable yields and environmental protection. Even so, many rural areas have suffered considerable degradation owing to excessive tree felling. Petroleum provides more than 80 per cent of all commercial energy and is the main source of energy for the transport, industrial and agricultural sectors. All petroleum is imported.

Increasing power generation is another key long-term consideration for Kenya. Total installed capacity for electricity amounts to 842 megawatts (MW): 71 per cent from hydro power, 20 per cent thermal, 5 per cent gas turbine, while some 4 per cent is imported from Uganda. The main operator here is the Kenya Power and Lighting Company, up to 40 per cent owned by the government and one of the few utilities listed on the Nairobi Stock Exchange. It has dominated power generation, both as a producer and distributor, since the 1960s. While the company is able to use its immense size and outreach to pool resources for the development of energy sources, its monopolistic power and political clout (made possible by the Electric Power Act) give it a price leadership that often is predatory. Consumption of electricity has been growing at 5.7 per cent per year since independence, while the development of electric power capacity has lagged far behind. This led to system overloading, resulting in power cuts and rationing. Other energy sources include wind, solar, ethanol, geothermal, coal and bio-gas.

Kenya has not lacked new initiatives to expand infrastructure services. A number of power projects and road extensions have been planned in the past decade but failed to take off owing to lack of resources. A decade of less than harmonious relations with the donor community has meant the delay of many projects. The recent rapprochement with the donors is bringing back on-stream a number of infrastructure projects, with a huge road sector rehabilitation programme already agreed with the World Bank. The recent period has also seen considerable private sector interest in the utility sector.

12.3 Infrastructure and enterprise development

Africa's slow growth has partly been blamed on its inferior rate of infrastructure capital accumulation. However, while African governments tried to close the gap in the 1960s and 1970s by expanding socio-economic infrastructure, there have been serious cases of non-performance in recent years owing to poor budgetary outlays towards expansion and rehabilitation (Lee and Anas, 1992). This has resulted in power failures, inadequate and poorly managed telephone services, chronic water shortages and roads which (due to poor maintenance) obstruct communication more than they facilitate it. It has been argued that in the early to medium stages of development the growth in the demand for basic infrastructure, including power generation, transport and communications, telecommunications, irrigation and agricultural extension services, often far exceeds the growth of the economy itself (Neill, 1996). Infrastructure bottlenecks are thus bound to arise, becoming a drag on economic growth. To overcome this, it is necessary to accumulate an excess of 'social overhead capital'. This would provide something of a general subsidy to producers, lowering costs of production across the board. For many countries in Africa, however, this has of course not been possible since infrastructure expansion has received low priority in the budgets of the crisis-ridden economies (Lesueur and Plane, 1994).

While the role played by infrastructure in economic development has attracted considerable attention, recent analyses have not produced clear-cut results on the size and nature of its effects (Ford and Poret, 1991; Munnell, 1992). A problem relates to the measurement of infrastructure services at the aggregate level. While liberally used in economic analysis, the term 'infrastructure' is far from a homogeneous conceptualisation of goods, services or production level. For example, though roads and airports provide transport services, they operate under quite different environments. Similarly, while water-engineering techniques might be common to irrigation services as well as to electricity generation, the two are only vaguely related. In applications, therefore, proxies for infrastructure, such as public sector capital expenditure (Aschauer, 1989), or a measurable part of infrastructure, such as road mileage, are used to represent total infrastructure (Talley, 1996). The problems of interpretation and generalisation are thus considerable.

The above overview assumes that the impact of infrastructure is factor neutral. However, seen as an amenity to producers and consumers, infrastructure provision has implications for production decisions as well as labour markets. On the production side, the size and quality of infrastructure provision affects variable as well as fixed costs. An increase in variable costs – arising, for example, from the impact of poor infrastructure on transport costs – reduces the scale of production, but also directly affects the firm's price-setting decisions. On the other hand, the effect on fixed costs changes the profit-maximising level of output. In responding to the relative price shifts caused by malfunctioning infrastructure, the factor intensities of firms are bound to change. Firms might become capital- or labour-intensive depending on their possibilities for factor substitution. Furthermore, since infrastructure availability impacts on profitability, and thus the long-term viability of the firm, its quality influences the firm's location decision. Firms will thus locate in regions or countries where 'certain conditional thresholds' with regard to infrastructure supply are met. This in turn influences the spatial distribution of domestic industrial activity, in effect determining the size of the industry as well as its employment capacity (Duffy-Deno and Eberts, 1991).

When infrastructure is a consumer amenity, its supply directly affects labour market behaviour and consumer welfare. Malfunctioning infrastructure, for example, raises both travel time and transport costs. Thus the private costs related to wage employment rise, along with the general cost of living. Labour supply then falls as the interregional migration or travel to the target region is reduced. The degree to which the reduction in labour supply affects firms in the target region depends on the extent to which infrastructure affects firm productivity (i.e., the extent to which the supply of infrastructure reduces the firms' variable costs: see Gillen, 1996).

The consumer and producer amenity effects are illustrated in Figure 12.1 (see Dalenberg and Partridge, 1997, p. 271). Labour market equilibrium is at *A*, with a market clearing wage rate, w_0, associated to the employment rate, E_0. The demand, $D(I_0)$, and supply, $S(I_0)$, for labour are shown to be functions of the available infrastructure. Under the assumption that infrastructure is an amenity to households, (i.e., reduces travel time and expenses), a decline in infrastructure supply pushes the labour supply

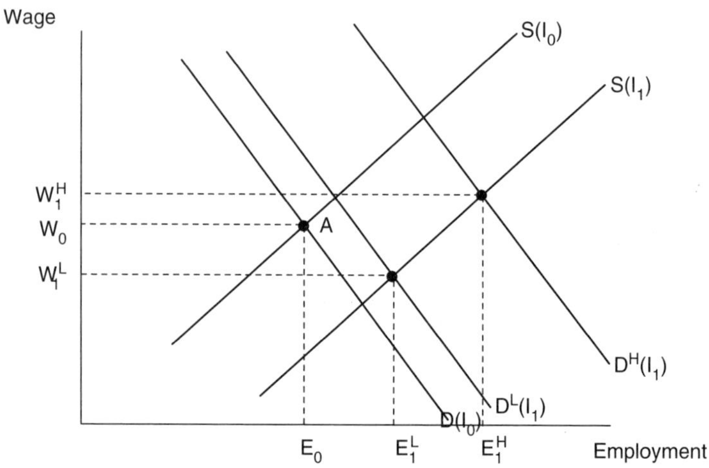

Figure 12.1 Employment and wages when infrastructure is a producer and consumer amenity

schedule to $S(I_1)$. Under the extreme assumption of infrastructure having no productivity effects on firms, its decline simply cuts labour supplies and raises wages (employment to E_1 and wages to w_1). However, if the productivity effects of infrastructure on production are high, then a reduction in infrastructure supply reduces both employment (labour demand schedule shifts to $D(I_1)$) and wages. In sum, while the impact on employment is unambiguously negative via the amenity and productivity effects, that on wages is unclear. Focusing only on the impacts of infrastructure on employment would make it difficult to evaluate the extent to which infrastructure is an amenity to firms (as opposed to households). The distinction is important because each alternative has distinct policy implications. Dalenberg and Partridge (1997) have argued that when the consumer (or worker) amenity effects dominate, workers will be attracted to regions or sectors with improved infrastructure, leading to falling wages. Alternately, if productivity effects dominate, firms will demand more workers and wages will have to rise (see also Holtz-Eakin and Lovely, 1996).

Generally, results from empirical estimations of the importance of infrastructure supply to economic performance have not been clear-cut. It has been suggested, therefore, that while there might be a common factor that binds growth and infrastructure together, it is difficult to capture it in the analysis. As has been argued in the case of aid and development (Burnside and Dollar, 1997), it might well be that infrastructure expansion is most effective in good policy environments but could be downright ineffective in bad ones. In African oil economies, for example, the rapid expansion in infrastructure in the 1970s, in the face of booming incomes, was poorly planned and the increased supply of infrastructure was not beneficial to the economy.

It also has been argued that infrastructure performs badly in economies where there is little competition in service provision. Lack of 'contestability' enables public utilities, operating under soft budgets, to underperform without market sanction (Baumol and Lee, 1991). Lack of private sector alternatives makes firms captive, forcing them to adjust capacities. In the face of deteriorating infrastructure, enterprises will either relocate, undertake factor substitution, reduce output or resort to self-provision (Departments of Economics, 1994). In the process, the response to infrastructure deficiencies revolves around the interplay between infrastructure supply, production technologies and markets for infrastructure aggregates. In power provision, for example, firms might seek total self-sufficiency or some self-provision while retaining an active interest in public provision. They could also choose to maintain stand-by facilities or be satellites with respect to power services (Lee and Anas, 1992). When markets for infrastructure are opened up, firms cease to be captive and a more reliable supply improves production and lowers costs (Gramlich, 1994).

These different regimes and their implications on firm production can be captured by specifying production functions that take into account the input of infrastructure services of varying quality and cost, depending on the nature of the firm's response. Modelling the production behaviour of an enterprise that is captive with respect to public infrastructure should thus incorporate the loss in output resulting from unanticipated input fluctuations: for example, due to power failure.[2] The behaviour of the self-sufficient firm can be captured in a nested optimisation set-up in which the firm seeks to maximise profits subject to costs. Self-provision alters the cost structure and, therefore, the point of optimality. Between these polar cases lie other potential responses to power supply related problems, such as purchase of stand-by facilities, or creation of satellite and joint production firms. However, whatever the response to deficient infrastructure might be, firms incur costs that are higher than if public sector provision were efficient.

12.4 Impact of the deterioration of infrastructure on firms

12.4.1 Correlation of infrastructure problems

Infrastructure problems are by no means isolated; they are dependent and actively reinforce each other. In Table 12.2 we present Pearson correlation coefficients between firm assessments of the extent of their problems with infrastructure provision. The coefficients are generally high and significant, but with some interesting differences. The correlation coefficient between roads and waste disposal is relatively high (0.43). This indicates the obvious fact that poor roads and congestion in the industrial areas have made access very difficult, especially for rubbish collection. For similar reasons, the correlation between telephones and roads is high because when roads are impassable, access for repairs and service becomes difficult. We also note that security and roads are significantly correlated, again relating to

Table 12.2 Correlations of infrastructure problems by type

	Electricity	*Water*	*Roads*	*Telephones*	*Waste*	*Security*
Electricity	1					
	(0.0)					
Water	0.2111	1				
	(0.0016)	(0.0)				
Roads	0.1632	0.1069	1			
	(0.0147)	(0.112)	(0.0)			
Telephones	0.2458	0.1088	0.3363	1		
	(0.0002)	(0.1059)	(0.0001)	(0.0)		
Waste	0.1329	0.1402	0.4258	0.1791	1	
	(0.0474)	(0.0368)	(0.0001)	(0.0073)	(0.0)	
Security	0.0288	0.0751	0.1906	0.1919	0.2221	1
	(0.6795)	(0.2834)	(0.0059)	(0.0056)	(0.0013)	(0.0)

access. It is noteworthy, however, that the waste disposal problem is correlated with all other infrastructure types, implying that it affects all firms. Water is only significantly correlated with waste disposal and electricity, while electricity is correlated with all infrastructure types except security. Table 12.2 thus illustrates cross-cutting effects, which are likely to affect the performance of manufacturing enterprises.

12.4.2 What enterprises consider as the greatest infrastructure problem

During the RPED interviews, firms were asked to indicate their greatest problem with respect to infrastructure provision. Given that infrastructure comprises disparate components, however, it is difficult to devise a measure that accounts for the total effect. Still, with regard to specific infrastructure types such as electricity and roads, the sample firms had a clear perception of the extent to which their activities were affected by the poor provision. Table 12.3 shows that in the early 1990s, roads had become the 'greatest' infrastructure problem for over 30 per cent of the formal firms, while electricity was a serious problem for about 20 per cent of the firms. Security was considered the 'greatest' problem by close to 18 per cent of respondents, while other infrastructure types had a much lower degree of severity, with less than 10 per cent considering telephones, water, and waste disposal to be among the greatest infrastructure problems they were facing.

As expected, there is considerable variation in responses depending on the sector and size of the firm. Thus while electricity provision is crucial for all firms, Table 12.3 shows that textile firms were much more affected by poor electricity provision than firms in other sectors. Similarly, water

Table 12.3 Greatest problem with infrastructure by type and sub-sector (%)

Sector	*Electricity*	*Water*	*Freight transport*	*Workers' transport*	*Roads*	*Telephones*	*Ports*	*Waste disposal*	*Security*
Formal									
Food	19.5	17	2.4	–	36.8	4.9	–	9.7	9.7
Textiles	28.6	8.6	–	–	28.6	11.4	–	11.4	11.4
Wood	9.8	7.3	–	–	51	4.9	–	–	24.4
Metal	24.3	2.7	5.4	–	16.2	13.5	2.7	10.8	24.3
Average									
Formal	20.1	9	2	–	33.15	8.7	0.6	9.9	17.5
Informal	36.8	3.5	1.8	5.3	14	10.5	–	8.8	15.8

Note: There was a category of 'other', which means that rows sometimes do not sum to 100%.
Source: Computed from the 1993 survey wave.

supply is more constraining in the food sector, where it is a crucial input throughout the production chain, than in others. On the other hand, only about 4 per cent of firms in the metal sector considered poor water provision as their greatest infrastructure problem.

A similar pattern emerges with respect to roads. The wood sector is transport-intensive, with the raw material either imported from neighbouring countries or collected from remote regions of the country. Furthermore, sawmills are located near forests, to ensure a steady and less costly supply of timber, and thus away from the main roads. Over 50 per cent of the firms in the wood sector saw roads as their greatest infrastructure problem, while, for similar reasons, the food sector had 37 per cent of the respondents indicating roads as a serious problem. The textile and metal sector also depend on bulk transport for their supplies and distribution of output. Roads were said to be a serious problem for up to 28 per cent of firms in these sectors.

In the early 1990s political conflicts related to demands for multiparty democracy and the elections increased the level of insecurity in Kenya. In the run-up to the national elections of 1992, for example, there were violent eruptions in various parts of the country. It would appear that the wood and metal sectors had the most serious problems with security, with close to a quarter of the firms seeing it as their greatest 'infrastructure' problem. This is perhaps related to the fact that the metal sector in Kenya is dominated by firms owned by Kenyans of Asian origin and by expatriate firms. Both groups have suffered economically, if not physically, in periods of political crisis. Though many firms in the sector were small and informal, there were large sawmills in politically destabilised regions such as Burnt Forest in the Rift Valley. These were forced to scale down or suspend operations owing to insecurity.

In Table 12.4, we look at the problems with infrastructure from the point of view of the size of the firm. Two observations can be made about the pattern arising in the table (compare this with Table 12.3). First, the impact on the smallest companies, referred to as the 'informal sector', seems to differ markedly from the rest. From their point of view, infrastructure can be divided into a part with 'free' access, notably roads, and a part with 'restricted' access, such as electricity, water and telephones. Access is restricted because firms need to satisfy certain requirements in order to have access to the relevant infrastructure. The requirements in Kenya include formal registration with the registrar of companies and possession of a physical business address. However, while access to infrastructure is a clear benefit of formality (visibility), there are also costs: when firms become formal, their liability to taxation increases.

With regard to roads, where access is free, informal firms are least affected. Their markets are often in close proximity to their business premises, while the size of the business does not demand large stocks or transport of bulky goods. On the other hand, the bigger companies are much more vulnerable

Table 12.4 Greatest problem with infrastructure in manufacturing by type and firm size, 1993 (%)

Size	*Electricity*	*Water*	*Freight transport*	*Workers' transport*	*Roads*	*Telephones*	*Ports*	*Waste disposal*	*Security*
Formal									
1–5	15.4	23	–	–	23	7.7	–	7.7	23
6–20	30.8	7.7	–	–	34.6	3.8	–	15.4	7.7
21–75	5.2	13.8	–	–	31	12	1.7	6.9	29
76–500	20	2.7	2.7	–	50.4	8	–	5.4	10.8
501 +	20	–	20	–	20	10	–	10	10
Average Formal	20.1	9	2	–	33.8	8.4	0.6	7.8	17.5
Average Informal	36.8	3.5	1.8	5.3	14	10.5	–	8.8	15.8

Note: Firm size is by number of employees.
There was a category of 'other', which means that rows sometimes did not sum to 100 per cent.

to poor road infrastructure. For example, over 30 per cent of medium to large firms saw roads as their greatest infrastructure problem.

With respect to the 'restricted' infrastructure, informal and smaller small firms considered electricity provision to be their greatest infrastructure problem. In their case, the problem extends beyond inadequate performance or power failures to lack of connection to the mains for reasons pointed out above. Table 12.4 shows that these problems were repeated for telephones. With respect to security, the very small firms are again more affected than average. They often operate in a grey zone where informality reduces their legitimacy and exposes them to higher levels of insecurity. Small firms also operate close to their clients, often as extensions of household activities, in the more densely populated sections of the urban centres. They are thus often embroiled in the political and civil disturbances. Frequently, their business activities are badly affected during police 'clean-up' operations.

As noted in the overview above, access to infrastructure supply can be problematic for three major reasons. First, while the requisite infrastructure might be in place, poor maintenance and inadequate capacities lead to frequent breakdowns and thus to deficient operation and inadequate supply of services. However, poor access might also simply a result of lack of infrastructure, as when the government fails to extend water supply to certain areas. Furthermore, the cost of infrastructure services might be prohibitive.

With the help of Table 12.5, we discuss briefly the firms' assessment of the quality of infrastructure provision in relation to the three constraints: deficient operation and poor supply of services, absence of the service, and prohibitive costs.

Table 12.5 indicates that deficient infrastructure was cited as the main cause of poor service provision by formal firms. As many as 86 per cent of

Table 12.5 Why infrastructure of a certain type is a problem (%)

	Electricity	*Water*	*Telephones*	*Roads*	*Waste disposal*
Formal					
Deficient	86.3	73.6	64.6	91	42.2
Not available	3.2	16.7	5.1	3	50
Too expensive	6.3	–	15.2	–	2.6
Other	4.2	9.7	15.1	6	5.2
Informal					
Deficient	37.1	37.5	12.5	88.5	33.3
Not available	34.3	43.8	50	7.7	52.4
Too expensive	8.6	–	16.7	–	–
Other	20	18.8	20.8	3.8	14.3

these firms cited electricity failures, including power dips, often occurring at peak production hours, as the main cause of disruption to production. A similarly large proportion (74 per cent) had problems with the breakdown of water supply. Over 60 per cent of the respondents reported frequent disruption of telephone services, while about 40 per cent saw disruption and breakdowns of the waste disposal system as a serious impediment to business. However, road transport was the most problematic, with 91 per cent of the sample reporting that roads were seriously deficient.

Still, owing to slow infrastructure expansion in the face of rapid population growth, poor access is often simply a result of lack of provision. Over 3 per cent of the respondents found electricity supply to be a problem because it was not provided at all; 5 per cent found it difficult to acquire telephone lines because they were not available; while over 17 per cent could not be connected to water for lack of meters and other equipment. With regard to waste disposal, 50 per cent of the respondents simply had nowhere to deposit their waste.

It is significant that formal firms do not consider the cost of infrastructure services *per se* as a major constraint. Only 6 and 15 per cent indicated that electricity and telephone tariffs, respectively, were too high. Ironically, the poor services are partly to blame on the government's reluctance to charge economic prices (that is, much higher prices).

With regard to security, a slightly different form of infrastructure from the rest, respondents saw the escalating cases of armed robbery as a serious problem. As noted earlier, the increase in insecurity was related to the political tension that arose during and after the election campaigns of 1992, and thus the escalation was perhaps of a one-off character. Still, burglaries were reported by many firms, which blamed the burglaries for the rising cost of security provision.

For informal firms as a whole, infrastructure was a problem mainly because it was not available. Thus, except for roads, informal firms had poor access to infrastructure.

In the 1994 wave, firms were asked if they had noted improvements in infrastructure provision by type since the interviews a year earlier. They were also asked to state where they had noted higher or lower profits in the previous year. In Table 12.6, firms are differentiated with respect to whether they had experienced higher or lower profits in the previous year.

On average, over 80 per cent of the formal firms noted deterioration in roads. The food and wood sectors, whose activities are transport-intensive, reported the largest deterioration. It is noteworthy that firms reporting lower profits also experienced a more serious deterioration of road services. Firms also experienced a sharp deterioration in telephone, water and electricity services. However, in the latter cases there does not seem to be a clear relationship between deteriorating service provision and changes in profits. It is obvious, however, that poor and deteriorating infrastructure sometimes coexisted with higher firm profits.[3] The discrepancy illustrates interesting features of firm adaptation: it is possible that in the short run firms can devise shock absorbers which can see them through the difficult infrastructure downturns. This would include self-provision of infrastructure, as noted elsewhere in this chapter. However, without an adequate infrastructure, markets are bound to remain thin and profits cannot be sustained. Poor infrastructure also is a form of entry barrier in some sectors. It definitely raises start-up costs, enabling veteran firms that adapted to bad conditions over the years (by, for example, acquiring their own generators and water tanks) to enjoy monopolistic profits when markets expanded.

Table 12.6 Deterioration in infrastructure by type and sector and change in profits, 1994 versus 1993 (%)

Sector	*Electricity*	*Water*	*Roads*	*Telephones*
Food (H)	56.5	47.8	78.3	60.9
Food (L)	33.3	41.6	91.7	66.7
Wood (H)	15.8	52.6	78.9	57.7
Wood (L)	26.7	53.3	93.3	64.3
Textiles (H)	62.5	50	81.3	62.5
Textiles (L)	16.7	45.5	75	63.6
Metal (H)	46.7	42.9	80	53.3
Metal (L)	61.5	53.8	76.9	61.5
Informal (H)	6.7	–	50	–
Informal (L)	29.6	26.9	46.9	15

Note: H = firm registered higher profits; L = registered lower profits.

12.5 Amenity effects of infrastructure provision

In this section we investigate the extent to which sector and region-specific differences impact on the provision of infrastructure. This investigation is done by regressing firm 'infrastructure scores' (with no problem on a particular infrastructure scoring 1, and serious problems scoring 5) on dummy variables for the region where the firm is situated, its sector of description, as well as on the size of the firm itself.

Following the theoretical aspects outlined above, we expected that regions which are at a remove from markets (the bigger urban areas) would be more dissatisfied with their road services, while those in the proximity of the cities would complain more about congestion or poor waste disposal. Likewise, it was expected that the metal sector might be more affected by poor electricity supply than the food sector, while the latter might be more affected by poor water supply. We were also interested to see what impact size has on the infrastructure scores for the various types. Since our size variable is based on total employment, the sign of the coefficient for size could give us some feel for the relationship between a firm's perception of the adequacy of infrastructure services and its level of employment.

Table 12.7 presents results from OLS regressions where the dependent variable is the 'firm score' for the various infrastructure types discussed above. The explanatory variables are dummies for Nairobi, Mombasa and Nakuru (with Eldoret as the reference), dummies for textile, wood and food sectors (with the metal sector as the reference) and the size of the firm proxied by total firm employment. A quick evaluation shows that the regressions for roads and air and sea transport had the 'best fit', although generally the results collaborate rather well with the discussion above. Still, there are a number of divergent results that need some elaboration.

In Table 12.7, a negative coefficient indicates that, compared to the reference region or sector, the relevant region or sector had fewer firms that have experienced problems with infrastructure of the given type. A negative sign for size indicates that the bigger the firm, the fewer problems it has with infrastructure.

In the regression for electricity, the dummy for Nairobi was significant, with a large coefficient. Relative to Eldoret, electricity provision is perceived by firms in Nairobi to be poorer than in other towns in the sample. With respect to water, Nakuru firms have more serious, and statistically significant, problems with water than Eldoret and other regions of the sample. The table shows that compared to the metal sector, the food sector has serious problems with water supply, with a relatively large and significant coefficient.

The results for freight transport and air and seaports are quite similar. With respect to freight transport, the textile sector seems to have fewer

Table 12.7 Estimation results for firm assessment of infrastructure problems

	Electricity	*Water*	*Freight transport*	*Roads*	*Telephones*	*Air and seaports*	*Waste disposal*	*Security*
Intercept	2.08***	1.66***	1.42***	1.56***	2.68***	1.4***	1.8***	3.1***
	(0.441)	(0.395)	(0.302)	(0.442)	(0.418)	(0.235)	(0.43)	(0.43)
Nairobi	0.68*	−0.073	0.175	1.46*	0.0525	0.291	0.727*	−0.2
	(0.412)	(0.369)	(0.282)	(0.413)	(0.3911)	(0.22)	(0.402)	(0.404)
Mombasa	0.445	0.461	−0.18	−0.218	−0.681	0.019	−0.252	−0.629
	(0.469)	(0.421)	(0.322)	(0.471)	(0.446)	(0.251)	(0.458)	(0.460)
Nakuru	−0.074	0.897*	0.235	0.902*	−0.279	0.312	−0.103	0.065
	(0.532)	(0.476)	(0.365)	(0.533)	(0.504)	(0.284)	(0.519)	(0.522)
Textiles	0.231	0.141	−0.373*	−0.087	−0.367	−0.488***	−0.141	−0.683**
	(0.318)	(0.285)	(0.218)	(0.319)	(0.302)	(0.170)	(0.311)	(0.312)
Wood	−0.224	0.017	0.055	0.928***	0.146	−0.423**	0.160	0.009
	(0.332)	(0.297)	(0.227)	(0.333)	(0.315)	(0.177)	(0.324)	(0.325)
Food	0.28	0.64**	0.271	0.251	−0.237	−0.423**	0.333	−0.614*
	(0.337)	(0.302)	(0.232)	(0.339)	(0.321)	(0.180)	(0.329)	(0.331)
Size	−0.026	0.025	0.038*	0.012	−0.0178	0.074***	−0.0437	0.029
	(0.031)	(0.028)	(0.021)	(0.031)	(0.030)	(0.017)	(0.031)	(0.031)
R^2	0.05	0.08	0.076	0.20	0.051	0.144	0.08	0.067
Prob > F	0.1913	0.021	0.028	0.0001	0.1607	0.0001	0.018	0.057

Note: Except for the size variable, which was proxied by total firm employment (divided by 100 to scale it down), the rest of the explanatory variables are dummies (standard errors are in brackets). Since we are using dummies, the reference for the regions is Eldoret and that for the sectors is metal. The asterisks (*, **, ***) indicate significance at the 90, 95 and 99 per cent levels of confidence.

problems than other sectors. The same seems to be the case with respect to air and seaports.[4] Relative to the metal sector, textiles, wood and food seem to have fewer problems with air and seaports. This is most probably because they are less dependent on imported inputs, and exports of finished products, than the metal sector. The wood and food manufacturing sectors get their supplies from domestic sources, while they export little or not at all. The textile sector also receives the bulk of its inputs from domestic sources.

With regard to roads, firms in Nairobi and Nakuru are relatively more affected than those in Eldoret or Mombasa. In Nairobi, firms also complained about the high level of congestion in the industrial areas of the city, especially since the roads' carrying capacities remain at the pre-independence levels while traffic has increased many times over. It is also significant that the dummy for the wood sector has a large and very significant coefficient, indicating the difficulties that lumbering and wood processing firms have in transporting bulky wood products over poor roads.

The outcomes for telephones were insignificant, but a number of the coefficients had negative signs (e.g., for Mombasa and Nakuru), showing that telephones were less of an impediment to activities than other infrastructure types. With respect to waste disposal, Nairobi seems to be the city where the problem is most significant, while the negative signs for Mombasa and Nakuru indicate a less serious problem there.

Regarding security, the problem seems to have been less pressing in Nairobi and Mombasa than in Eldoret and Nakuru. At the sector level, textiles and food have fewer problems with security than the other sectors.[5]

Finally, let us revisit the size variable. As indicated earlier, the size of the firm was simply inferred from the number of its employees. The size variable is only significant in two cases, freight transport, and air and seaports. Since the coefficients are positive, this also indicates that the bigger firms are more dissatisfied with the level of provision of these two categories of services than smaller firms. Better provision would lower firm costs and help preserve or increase employment. The two infrastructure types would in this regard be producer amenities.

Still, as we showed above, firm evaluations of the quality of infrastructure services are quite closely correlated and firms demand or need a good and adequate combination of services in order to function properly. It is clear that while a good supply of a particular infrastructure type could lower costs, inadequate supply of another could raise costs even more.

12.6 The determinants of self-provision

In order to remain in business, manufacturing enterprises in Kenya have been forced to address a wide range of infrastructure problems on their own. Ability to do this has varied with the size of the firm, as well as the sector in which it operates. The cost of self-provisioning as well as the potential returns of installing their own aggregates are also important considerations. In Table 12.8 we present the incidence of self-provisioning by sector, using data pooled from all three waves. While we showed above that poor roads and deficient electricity supplies are among the greatest infrastructure constraints facing Kenyan firms, they are not easy to remedy

Table 12.8 Self-provision of infrastructure services by sector (%)

	Formal					
	Food	*Textiles*	*Wood*	*Metal*	*Average*	*Informal*
Generator	14.9	5.3	6	10.5	9.17	0
Security	72.4	51	62.1	52.6	59.5	33.8
Waste disposal	32.8	23.7	28.4	23.7	27.2	8.9
Wells/cisterns	18.8	15.4	7.7	7.9	12.5	0.6
Transport						
workers	11.8	7.3	7.7	6.8	8.4	1.8
freight	19.2	5.6	18.1	5.2	12	0

Table 12.9 Self-provisions of infrastructure services by firm size (%)

	Formal						
	Micro	*Small*	*Medium*	*Large*	*Very large*	*Average*	*Informal*
Generator	0	1.9	10.2	16.8	22.7	9.27	0
Security	23.1	45.3	70.5	75	60	59.5	33.8
Waste disposal	10	22.7	36.7	29.1	45.4	27.2	8.9
Wells/cisterns	5	5.6	15.3	18.4	40.9	12.6	0.6
Transport							
workers	5	8.5	6.8	12.9	31.8	8.3	1.8
freight	0	9.4	14.2	16.9	40	12	0

at the private level. Thus only about 9 per cent of the formal firms were able to install a power generator, while none in the informal sector was able to do so. Indeed Table 12.9 shows that larger firms were more predisposed to purchasing generators than smaller ones. Close to 20 per cent of the larger companies had installed private power generators.

The acquisition of a regular source of power seems closely connected to the pursuit of other corporate objectives. Thus more than half of the enterprises that acquired stand-by generators also bought some other new equipment Kimuyu, 1998a, 1998b). However, save for cases in the urban areas, such as in Nairobi's River Road where several informal producers had purchased a reconditioned generator for joint use, few informal or small firms purchased new standby equipment. The high cost of the power aggregates is a major deterrent. Thus, on the whole, informal or micro enterprises were more likely to be captive to public utilities than their formal sector counterparts.

Besides a well-developed infrastructure, peace and security were key factors for choosing Kenya as a business location. However, the past decade has seen a marked deterioration in the security situation. In a bid to protect their investment, firms responded by taking increasing recourse to private security arrangements. Table 12.8 shows that close to 60 per cent of the firms had hired security guards or made other security arrangements. The food and wood sectors seem to have had a higher need for security than other sectors. Further, firms with their own security arrangements are larger than average (Kimuyu, 1998b). However, while less than 50 per cent of the smaller firms make security arrangements, the corresponding figure for medium and larger firms is over 60 per cent. As in the case of power generation above, enterprises with their own security provisions also seem to have increased investment in equipment (Kimuyu, 1998a). This again reflects that business decisions are closely interwoven with availability of infrastructure. In this case infrastructure availability is driving investment. A spin-off from the increasing need for security has been the rapid growth of a market for the provision of security services in Kenya.

An expansion of manufacturing activities also means increased amounts of industrial and other waste. However, it is now clear that the industrial areas put aside during the colonial era have been overcrowded, accommodating more companies than can be provided with an adequate infrastructure. Table 12.8 shows that about a quarter of the firms had made their own arrangements for waste disposal. However, this in no way implies that the rest were adequately catered for. Indeed the dumping of waste is more the rule than the exception in parts of Nairobi's Industrial Area. Table 12.9 again shows that the larger companies were more disposed to making private arrangements for waste disposal than smaller ones.

To counter the unreliable public services, enterprises have put in place their own water systems, usually in the form of boreholes and water cisterns. The tables show that over 10 per cent of the formal firms had standby arrangements for water supply, with larger companies more inclined to have such arrangements for fear of the disruptive impacts of supply interruptions. It is again indicative that more than half of the enterprises that acquired new equipment also had a standby source of water supply.

Firms are constrained by the availability of transport for workers, as well as freight transport for inputs and finished products. Tables 12.8 and 12.9 indicate that only 8 per cent of the firms provided transport for their workers. Again the larger companies are best placed to do so, with over 30 per cent of the largest category providing transport for their workers, while only 5 per cent of the smallest firms were able to do so. Similar trends are evident in the case of freight transport. About 10 per cent of the firms provide their own freight transport, with larger companies doing it more frequently than smaller ones. Close to 40 per cent of the larger companies had access to their own freight transport, but none of the smaller ones.

Table 12.10 presents results from a probit analysis of the determinants of self-provision of infrastructure services. The explanatory variables are firm size, proxied by the number of employees, engagement in export activities (we use a dummy for exporting), business location and sector affiliation.

The results suggest that firm size and engagement in export activities raise the probability for the self-provision of most infrastructure services. Large operations need to be worked intensively, often on 24-hour shifts, to lower unit costs. Thus, as observed earlier, large firms, more often than smaller ones, resolve the problem of deficient infrastructure by purchasing their own generators, water tanks or creating dumps for waste disposal. Engagement in export activities also has a positive impact on the probability for self-provisioning. This is because the export market is very competitive so that failure to meet deadlines leads to loss of markets. Having their own infrastructure reduces the risk of a total breakdown in production and thus ensures continued deliveries. Still, in a recent survey by the Kenya Association of Manufacturers (1996), 10 per cent of exporting firms had lost markets as a result of deficient infrastructure and an inability to replace it with a private alternative.[6]

Table 12.10 Probit results for the determinants of the self-provision of infrastructure

	Generator	*Water*	*Workers' transport*	*Freight*	*Waste*	*Security*
Intercept	−1.739*	−1.415*	−2.31	−1.249*	−0.751*	0.072
	(0.328)	(0.279)	(0.518)	(0.253)	(0.215)	(0.198)
Size	0.0347*	0.038*	0.107*	0.057*	0.034*	0.047*
	(0.020)	(0.019)	(0.024)	(0.021)	(0.020)	(0.027)
Exporter	0.761*	0.703*	−0.197	0.220	0.362*	0.420*
	(0.179)	(0.163)	(0.208)	(0.172)	(0.137)	(0.135)
Location dummies						
Nairobi	−0.136	−0.247	0.963*	−0.296	0.059	−0.166
	(0.313)	(0.264)	(0.502)	(0.231)	(0.203)	(0.189)
Mombasa	0.023	0.023	0.601	−0.648*	−0.623*	−0.618*
	(0.348)	(0.292)	(0.544)	(0.299)	(0.251)	(0.215)
Nakuru	0.006	0.076	0.568	−0.180	−0.342	−0.287
	(0.385)	(0.319)	(0.582)	(0.296)	(0.269)	(0.237)
Sector dummies						
Food	0.1605	0.251	−0.290	−0.033	0.059	0.219
	(0.214)	(0.186)	(0.236)	(0.212)	(0.155)	(0.142)
Textiles	−0.198	−0.299	−0.421*	0.233	−0.094	0.075
	(0.242)	(0.218)	(0.246)	(0.197)	(0.156)	(0.140)
Wood	0.094	0.097	−0.024	0.302	−0.034	0.088
	(0.221)	(0.197)	(0.215)	(0.198)	(0.159)	(0.145)
Log-likelihood	−148.76	−199.61	−337.90	−430.56	−199.88	134.12

Notes: Figures in brackets are standard errors. Eldoret and metal-working are excluded from the location and sector dummies. Figures marked with an asterisk are significant at the 10% level or better.

With respect to location, the results were generally insignificant, indicating that infrastructure problems were not confined to a particular location. Still, with regard to workers' transport, the Nairobi dummy was significant, indicating that compared to Eldoret, firms in the capital city were more likely to provide transport for their workers. Results for Mombasa indicate that firms situated there are less likely to self-provision in waste disposal, security, and freight transport. These results are consistent with Mombasa's less crowded industrial areas and better access to infrastructure services. Notably, being a port, freight transport is not as cumbersome as in upcountry Nairobi. Finally, though generally insignificant, the signs of the sector results indicate that, compared to other sectors, the probability for self-provisioning in textiles is lower. Most of the textile firms are small so that self-provision may not be feasible on account of scale diseconomies (see also Ndiege, 1998).

12.7 Conclusion

In this chapter, we have looked at the impact of Kenya's increasingly deficient infrastructure on the manufacturing sector. Though the country's infrastructure problems have been accumulating for most of the past two decades, as budgets for repair and expansion have declined, their seriousness became a major policy concern only in the era of structural adjustment. The market access and opening-up of the economy sought by the economic reforms might be difficult to achieve in the absence of an inefficient infrastructure.

The inability of the public sector to provide an adequate level of services has forced individual firms to devise their own coping mechanisms. Thus many have sought to change production techniques or to purchase their own infrastructure aggregates. However, although self-provisioning ensures a more regular supply of services, it often implies higher production costs. Moreover, in some cases, such as roads, railways, harbours and telephones, self-provisioning has in the past not been possible. Thus on balance firms were never entirely able to safeguard themselves from the effects of a malfunctioning infrastructure.

How, then, should the provision of services be enhanced? To bring about a real improvement, the flow of resources to infrastructure and related services has to improve. In recent years, the government has tried to do this, as in the case of the Road Maintenance Levy Fund, by increasing levies on automotive fuels. However, tax-based solutions without an underlying impact analysis do not always have a favourable effect on economic activity. In any case, it is necessary to devise broader forms of delivery, such as those based on a partnership between the government and the private sector, or the private sector and affected communities or sectors.

In searching for the means to improve infrastructure provision in Kenya, a number of important issues arise. First, there is needed to create an environment that is supportive of private sector participation. The appropriate legal structures, such as stipulations on power and water tariffs that enable viable private operations, need to be put in place. However, a number of groups benefited from the earlier controls and payment systems in the utility sector and the government needs to overcome their resistance, while focusing on an expeditious implementation of the new regulations.

Second, to increase efficiency in the delivery of services, individual consumers and companies must be given an exit option. It must be possible to punish poor service by moving to another supplier and liquidating poorly performing service suppliers. In the telecommunications sector, for example, services have been provided through non-transferable contracts between the parastatal and the consumers. With captive consumers, the parastatal responded slowly or not at all to complaints regarding poor service. It maintained large overheads and failed to be innovative, especially with regard to the introduction of new technologies. It is thus necessary to have more telecommunications carriers in order to encourage competition. This will lead to price discipline and reduced costs for consumers. The debut of a second mobile telephone service provider confirms this expectation. Similarly, the power industry in Kenya also needs more producers and independent distributors. As in the case of telecommunications, power consumers also need an unencumbered exit option.

Third, since the main responsibility for infrastructure provision is still in government hands, there is a need to devise ways of reducing the public sector's revenue constraint by introducing cost-sharing schemes. The argument is that if assured of improvements in service provision, communities are willing to meet a substantial portion of the costs of service delivery. In practice, however, cost sharing has proved difficult to implement at the poorer end of the income distribution. In poor urban areas and the rural sector, for example, water is often supplied communally and free-rider problems have been formidable. Still, involving communities in the planning and financing of water supply is one of the best ways of building systems that function and are sustainable (see Rondinelli, 1991).

Fourth, to provide infrastructure services at reasonable cost demands huge investments in production capacities. This is often an effective entry barrier in poor countries. Aggregates, (in the power sector, for example) are difficult to unbundle, and during privatisation are likely to be purchased by a single private operator. This in turn confers a degree of monopoly power on private utility companies which must be countered by a national regulator. However, the activities of the national regulator need to be clearly specified in order to prevent it from falling back on the control tactics of old. It will have to ensure that standards are upheld, but without being obtrusive. Such an agency would also assume the planning and development function for the relevant infrastructure service. It will need to take into consideration the long-term needs of the manufacturing sector and find ways in which the private sector could participate beneficially in service expansion and development. These arguments are pointers to possible terms of reference for the newly-created electricity regulatory board.

In conclusion, the government was the sole supplier of infrastructure in the past, and there might still be arguments in favour of continued state intervention in some areas, if only to ensure that the poorer urban areas and the rural sector have access to services. However, in the past the goals of a more egalitarian distribution of services (e.g., via urban upgrading, rural road construction and rural electrification) were defeated by paucity of resources and poor planning. There was a tendency to see utilities as a means of expanding modern sector employment. Little thought was given to the development of technical and management capacities. The government has now found it necessary to incorporate communities and the private sector in the search for ways to raise efficiency and improve service provision. The new situation will demand rapid institutional change, including new legislation to cater for the various aspects of the market for utilities in the wake of the dismantling of the state monopolies and the introduction of new technologies.

Notes

1. Some of these include the Ministry of Transport and Communications, Ministry of Public Works and Housing, local governments and the Ministry of Environment and Natural Resources.
2. For the captive firm, a possible formal presentation of its production configuration can be expressed as: maximise $\pi = P_q(1 - \sigma)F(I,\mu,X,E) - P_x X - P_e E$ where the choice variables are X and E representing factors of production and electricity, and where P_q, P_x and P_e are prices of output, inputs other than electricity, and electricity, σ is the failure rate of publicly supplied power, and μ represents the incidence of voltage fluctuations. In this formulation, it is assumed that power failure is randomly distributed and that production is continuous so that the enterprise ends up paying for the idle time since it is unable to anticipate the frequency and duration of random power failures. For the self-provider, there is no 'uncontrolled' risk of failure such as that entailed by reliance on poor public infrastructure, although unit costs are correspondingly higher.
3. Figures drawn from the third wave (1995) show that firms which thought that profits had risen were fewer in number than in the previous year, but still substantial, while infrastructure provision had, according to the firms, deteriorated further.
4. It is, however, possible that the definition of the textile sector used here groups together tailors and other individuals working in the informal sector with the bigger textile companies. The smaller firms have a ready market and might not have freight transport problems.
5. As noted above, the questions related to security had more immediate implications in the region of the Rift Valley, where Nakuru and Eldoret are situated, than elsewhere in the country during the aftermath of the 1992 elections.
6. An example is given of Bamburi Portland Cement Company, an exporter, which reported a loss of 50 000 tonnes of cement as a result of power interruptions in 1994 (Kenya Association of Manufacturers, 1996, p. 33).

References

Aschauer, D.A. (1989), 'Is Public Expenditure Productive?', *Journal of Monetary Economics* 23(2).

Baumol, W.F. and Lee, K.S. (1991) Confestable Markets, Trade and Development', *World Bank Research Observer* 6(1).

Burnside, A. and Dollar, D. (1997), 'Aid, Policies and Growth', Policy Research Working Paper No 1777, World Bank, Washington, DC.

Dalenberg R.D. and Partridge, M.D. (1997), 'Public Infrastructure and Wages', *Land Economics* 73(2).

Duffy-Deno, K.T. and Eberts, R.W. (1991), 'Public Infrastructure and Regional Economic Development: A Simultaneous Equations Approach', *Journal of Urban Economics* 30.

Economics Departments, University of Nairobi and Göteborg University (1994), 'Limitations and Rewards in Kenya's Manufacturing Sector: A Study of Enterprise Development', Report to the World Bank, Washington, DC.

Ford, R. and Poret, P. (1991), 'Infrastructure and Private-Sector Productivity', *OECD Economic Studies* 17.

Gillen, D.W. (1996), 'Transportation Infrastructure and Economic Development: A Review of Recent Literature', *Logistics and Transportation Review* 32(1).

Gramlich, E.M. (1994), 'Infrastructure Investment: A Review Essay', *Journal of Economic Literature* 32(3).

Holtz-Eakin, D. and Lovely, M.E. (1996), 'Returns to Variety, and the Productivity of Public Infrastructure', *Regional Science and Urban Economics* 26(2).

Kenya Association of Manufacturers (1996), 'Private Sector Involvement in Development, Operations and Maintenance of Kenya's Infrastructure', Consultants' Report, KAM, Nairobi.

Kimuyu P.K. (1998a), 'Equipment Acquisitions by Kenyan Manufacturers', Discussion Paper No. 9, Institute of Policy Analysis and Research, Nairobi.

Kimuyu P.K. (1998b), 'Industrial Policies for the Twenty-First Century: Productivity, Competitiveness and Export Participation by Manufacturing Enterprises in Kenya', Paper presented at the Institute of Policy Analysis and Research Annual Workshop on Strategic Development Paths for Kenya in the 21st Century, Nairobi, 15–16 April.

Lee, K.S. and Anas, A. (1992), 'Costs of Deficient Infrastructure: The Case of Nigerian Manufacturing', *Urban Studies* 29(7).

Lesueur, Jean-Yves and Plane, P. (1994), 'Human Resource Management and the Restructuring of Public Utilities: Water and Electricity in Africa', *International Labour Review* 133(3).

Munnell, A. (1992), 'Infrastructure Investment and Economic growth', *Journal of Economic Perspectives* 6(4).

Nalo, D.S.O. (1993), 'Adequate Road Maintenance in Kenya: The Need for Appropriate Policies', *International Journal of Transport Economics* 20(2).

Ndiege P. (1998), 'Increasing Access to Energy by Micro and Small Scale Enterprises in Kenya', Paper presented at an International Centre for Economic Growth seminar on 'Improving Access to Infrastructure by Micro and Small Scale Enterprise', Nairobi, 9 March 1998.

Neill, J.R. (1996), 'Fuelling the Engine of Growth with Investment in Infrastructure: A Lesson From Neoclassical Growth Theory', *Journal of Macroeconomics* 18(3).

Republic of Kenya (1994), *Sessional Paper No. 1 of 1994 on Recovery and Sustainable Development in the Year 2000*, Government Printer, Nairobi.

Republic of Kenya (1995), *Sixth Report of the Public Investments Committee on the Accounts of State Corporations*, Vol. 1, Government Printer, Nairobi.

Rondinelli, D.A. (1991), 'Decentralizing Water Supply services in Developing Countries: Factors Affecting the Success of Community Management', *Public Administration and Development* 11(5).

Talley, W.K. (1996), 'Linkages Between Transportation Infrastructure Investment and Economic Production', *Logistics and Transportation Review* 32(1).

Wheeler, D. and Mody, A. (1992), 'International Investment Location Decisions: The Case of U.S. Firms', *Journal of International Economics* 33(1–2).

World Bank (1990), *Kenya: Urban Transport Development Issues*, Report No. 7881-KE, Washington DC.

13
Conclusions

Arne Bigsten and Peter Kimuyu

Industrial development in the world in recent years has been characterised by a rapid globalisation of production, facilitated by liberalisation of international economic relations and changes in the character of corporate activity and international investment. As a consequence, production systems have become increasing integrated across national borders. Firms worldwide have been exposed to increased competitive pressure, and in large parts of the world production and international trade have grown rapidly. Countries in Africa have been part of this phenomenon of liberalisation and opening-up, but here the process has been slower and less pervasive.

Africa's share in global manufacturing production was a tiny 0.3 per cent in 1995 (UNIDO, 1996) which represents a fall from 0.4 per cent 10 years earlier. Even in labour-intensive production, the African share is small. For wearing apparel, for example, Africa's share is only 0.6 per cent. This represents a small increase from 0.4 per cent in 1985. For capital-intensive manufacturing, Africa's share of global production is minute.[1] In summary, Africa accounts for a very small part of world manufacturing output. Changes in production patterns since 1985 suggest that the continent has some comparative advantage in labour-intensive manufacturing.[2]

Kenya experienced a relatively fast rate of growth in the years immediately after independence, but from the mid-1970s the economy was hit by a series of shocks that revealed the weaknesses of the import-substitution strategy and made reforms necessary. The reforms that started in the early 1980s were initially slow and half-hearted. The government was reluctant to let go of its control of the economy, which was also a tool of political control. In the 1990s, however, there was finally an acceleration of the reform process, and at the end of our survey period in 1995 the economy had been very significantly liberalised. The system of government, though, has been slow to change, and the lack of good governance remains a crucial constraint on growth in Kenya. It is important to have good rules and regulations, but when the application of those is biased the development outcomes may still be poor. It is noteworthy that Tybout (1998, p. 42) concludes his review of the performance of manufacturing firms in less-developed countries with the observation that 'uncertainty about policies and demand conditions, poor rule of laws, and corruption may be the priority areas for reform'. This conclusion is certainly highly relevant for the Kenya case.

One ambition of the new, independent Kenyan government was to Kenyanise the economy. With regard to manufacturing there has been a considerable change in the pattern of ownership, with a dramatic increase in African participation. However, there are still substantial differences between African and Asian entrepreneurs. African manufacturers dominate the smaller, less productive and younger firms, which are often sole proprietorships, owner-managed and more diversified, and have less access to credit and operate in a more competitive environment. This contrasts with the larger, mostly Asian-owned, more specialised firms that exert more market power, employ specialised management and have a longer life expectancy. Foreign ownership is limited in Kenya, reflecting the fact that the country has not been able to create an environment that is attractive enough for foreign investors.

Large Kenyan firms have until recently been highly protected from competition and were oriented towards the domestic market, or possibly the regional market. Pressures for productivity improvements have been weak. At the time of the survey, the opening-up of the economy had only modest effects on the competitive situation of larger firms. Firms at the smaller and informal end of the spectrum, though, often work in a more competitive environment, but their ability to improve productivity is constrained by a host of other factors.

Except in the metal-working sector, the acquisition and use of technology licences and technical assistance is rare. Although the evidence for the productivity benefit of foreign equity participation is inconclusive, technological and efficiency benefits should accrue from such participation. Manufacturing firms in Kenya have yet to become well integrated into the world economy and the growth of manufacturing production has also been modest. Successful expansion of the industrial activities and export promotion will require facilitation of firm-level acquisition and use of licences and assistance.

The skill level in Kenyan manufacturing is rather low. The situation seems particularly acute with regard to technical skills, which is a reflection of the low level of technically-oriented education in Kenya. Although the general level of education of the population Kenya has been better than that in many countries in the region, there has been a stagnation or even a decline. To break into the world market and maintain participation, the country must develop its human resources by restoring school enrolments and completion rates (Abagi, 1998), and by upgrading technical education.

There are large size–wage differentials in the Kenyan manufacturing industry, suggesting deviation of the manufacturing sector's labour market from the competitive model. The large firms seem to pay high wages to attract and retain qualified and experienced workers. Higher wages also result from active internal markets within the large firms.

Although the financial markets have been reformed, problems stil abound. A weak legal framework, poorly defined property rights and imperfections in the financial markets are evident. These need to be addressed to increase the flow of financial resources and investment in the sector.

Investment levels were low during the period of study. Firms relied to a small extent on external sources of funds for investment. It does not seem to be the case that investment is constrained by lack of access to credit, but rather by the high costs of external capital or increased bankruptcy risk following formal borrowing. There is also evidence that equipment is difficult to sell at a reasonable price once installed. This forces firms that face a high sunk cost either to save the money or to place it in liquid assets for precautionary reasons. Our analysis shows that cash-flow does matter, but not as much as expected. Productivity seems to be a more important determinant of investment, suggesting that firms which have profitable investment opportunities do invest more than those that do not. Improved access to capital by itself would therefore have a limited effect on investment, unless accompanied by measures increasing productivity of investments. This brings us back to the array of factors that hamper economic development in Kenya.

Manufacturing firms are heterogeneous and general conclusions are difficult to draw. Still, productivity analysis reveals that food and wood-working sectors are more productive. From a productivity perspective, textile firms are somewhat peculiar in that they exhibit decreasing returns to scale. This peculiarity is related to the heavy reform costs borne by the textile sector. There is a strong, positive efficiency–size relationship so that larger firms are more efficient. Contracting firms are less efficient, so that overall productivity can be increased by accelerating firm growth and turnover. Access to credit has a greater impact on productivity than other policy variables, so that reforms which improve access to credit would have significant effects.

However, there is further concern about lack of product demand, presumably for firms with a narrow domestic focus. This is hardly surprising, considering the poor growth record of the country. Even with better performance, the domestic market is unlikely to sustain the desired growth rates, so that the case for an outward orientation is very strong. The potential benefits from export of manufactures are many: promotion of international competitiveness, stronger prop for international growth, concentration of investment in efficient sectors, promotion of specialisation, exploitation of scale economies, and increasing technical efficiency.

The decision by firms to participate in the export markets is a two-stage process, the first stage being related to a decision as to whether or not to export, followed by a decision about the export share of total production. Export-related risks reduce the attractiveness of specialisation, so export promotion is likely to be more successful where risks such as those related to exchange rate uncertainties and credibility of export programmes are reduced. Both sunk costs and expected profitability are important determinants of export decisions, and not surprisingly, Kenya's comparative advantage resides in labour-intensive products. Our data, however, offered only weak support for the export penetration hypothesis.

Although reforms have been beset by problems and policy reversals, the liberalisation process has improved availability of foreign exchange, imported raw materials, and the opportunity to import equipment. The reforms also opened up export opportunities. However, the contractionary policies instituted to stabilise the economy after the excesses during the 1992 election campaign were felt by the firms during the period covered by our interviews. The main complaints of the firms related to lack of credit, low demand, poor and expensive infrastructure and poor business support services. Still, the firms were often optimistic about their longer-term prospects, suggesting that they have some confidence in the adjustment programme. The adjustment process is less than complete, but there is evidence of some recovery, albeit fragile due to political uncertainty. Extensive domestic government borrowing has also crowded out the industrial sector. It will not, therefore, be possible to unfetter manufacturing activities without controlling the government's demand for domestic credit.

Deteriorating infrastructure and escalating utility costs hold back competitive production. Firm-level responses to deteriorating infrastructure improves stability of supply, but it also increases production costs, undermining the potential for the output expansion. Better results will be achieved by promoting infrastructure-related partnership between not just the government and the private sector, but between different private actors and affected communities. Simultaneously, industrial policy could also create exit options to punish poor service delivery by encouraging, where appropriate, secondary markets in infrastructure services. Policy will also have to direct revenue-improvement efforts towards selective introduction of cost-sharing.

Problems with enforcement of contracts make trade disputes common. The machinery for dispute settlement is inefficient, and parties to such disputes are forced to find *ad hoc* solutions outside the formal system where costs are low. Many of the disputes remain unresolved and, when they are, parties are dissatisfied with the outcomes.

Although many barriers to trade and investment have been dismantled, the response by investors has been limited.[3] In addition to poor institutions and infrastructure, the government has not been a consistent reformer, and its governance and transparency record is wanting. More needs to be done in reforming economic policies, but it is even more important to ensure effective and incorrupt implementation of policies once they are in place.

The regulatory environment in which Kenyan manufacturers operate has improved. Foreign exchange and price controls have been removed, and tariffs and taxes rationalised. Government controls are therefore less of a concern to entrepreneurs than they used to be. However, many still complain about deteriorating infrastructure. Although some firms have responded by self-provision for some of the infrastructure services, these strategies are sub-optimal as they reduce the country's export competitiveness and erode gains from a more liberal foreign exchange regime and trade reforms.

That aside, Kenya's industrial base can be expanded by creating conditions for increasing the life expectancy of small African-owned enterprises and reducing the risks and other circumstances that condemn firms to remain small, keep within the sole-proprietor form of ownership and stay within the low productivity trap. Reducing exit and entry barriers will also help to reduce market power and increase beneficial competitive pressure in the domestic market.

Notes

1. For chemical and transport equipment, for example, Africa's share of global production is estimated to be a mere 0.1 per cent.
2. Note, however, that Wood (1994) argues that Africa typically has a comparative advantage in natural resource-intensive sectors, while manufacturing is skill-intensive.
3. Sub-Saharan Africa's share in world foreign direct investment was 1.7 per cent in 1985, 1.0 per cent in 1990 and 1.3 per cent in 1995 (UNIDO, 1996, p. 49).

References

Abagi, O. (1998), *Education for the Next Millennium: A Demarcation of Imperatives*, Paper presented at the Institute of Policy Analysis and Research Annual Conference on Strategic Development Paths for Kenya in the 21st Century, Nairobi.

Tybout, J. (1998), *Manufacturing Firms in Developing Countries: How Well Do They Do, and Why?*, World Bank Research Working Paper, No.

UNIDO (1996), *Industrial Development. Global Report 1996*, Oxford University Press, Oxford.

Wood, A. (1994), *North-South Trade, Employment and Inequality: Changing Fortunes in a Skill-Driven World*, Clarendon Press, Oxford.

Index